Adsorption
from Solutions
of Non-Electrolytes

Adsorption
from Solutions
of Non-Electrolytes

J. J. KIPLING

Department of Chemistry
The University of Hull
England

ACADEMIC PRESS 1965

LONDON · NEW YORK

ACADEMIC PRESS INC. (LONDON) LTD
Berkeley Square House
Berkeley Square
London, W.1

U.S. Edition published by
ACADEMIC PRESS INC.
111 Fifth Avenue
New York, New York 10003

Library of Congress Catalog Card Number: 65-26421

CHEMISTRY

Printed in Great Britain by
The Whitefriars Press Ltd., London and Tonbridge

PREFACE

Adsorption from solutions of non-electrolytes is given little or no attention in current textbooks of physical chemistry. This may seem surprising in view of the many empirical uses which have been made of the phenomenon in chemical technology over many decades, notably in the purification of liquids, the separation of solutes, and the stabilization of dispersions (e.g. paints, printing ink).

The state of the literature suggests a probable explanation. As can be seen from the two volumes of Deitz's extremely useful *Bibliography of Solid Adsorbents*, the large number of research papers on the solid-liquid interface consists of isolated groups of two or three. Many authors have carried out one investigation of adsorption from solution and have then turned their attention elsewhere. Very few have spent enough time in the field to be able to reflect on its development and so produce a synoptic account of the subject. Without such general reviews or even lectures, the writers of textbooks have had no firm starting-point. Nor could they, as has been possible with other branches of physical chemistry, associate the early investigations with one of the great names of the classical period; adsorption from solution has had no Faraday, Arrhenius, or van der Waals. (It is worth pondering the question why we are now learning more from the mathematician Gibbs than from the experimentalist Freundlich.)

I have argued elsewhere that the early experiments did not provide the best foundation for building a theory of adsorption. Naturally enough, dilute aqueous solutions were used. This focused attention on the solute, and the solvent was ignored—a cardinal error when more concentrated solutions eventually came to be studied. Moreover, water (all too popular as a solvent) is not only quite untypical of solvents generally, but also gives more complex solutions than almost any other solvent.

The advent of World War I increased interest in the adsorption of gases by solids, and many chemists, concerned only incidentally with adsorption from solution, overlooked the difference between adsorption of a one-component gas and a two- (or multi-) component liquid. Slavish use of the Freundlich and Langmuir equations for adsorption isotherms has almost certainly done more harm than good to the theory of adsorption from solution.

In the last few years, however, more consistent attention has been given to the subject in a number of laboratories, which are trying to co-ordinate their results in spite of difficulties arising from distance, language, and international politics. Enough has now been done to show the probable scope of the subject. I have therefore judged it timely to attempt to sketch an outline map. Some of the detail has also been filled in; much, of course,

v

remains to be investigated. I hope that the sketch-map will direct attention to the areas where more attention is needed.

A general view of the subject is particularly needed at a time when we wish to see how far it will be possible to use theories of solutions to explain the phenomena of adsorption. The outcome of the present attempts is far from certain. Theories can only deal with very simple systems unless the equations are to become so elaborate as to be difficult or impossible to use. Experimentally, the more ideal a solution becomes, the more difficult it is to obtain accurate analytical results. One can still only speculate on the chances of a fully quantitative theory being developed. At least it can be claimed, however, that we now have a realistic model of adsorption from solution to serve as a basis of academic research. I hope that as it is discussed in relation to technological applications, it will not only lead to improvements in traditional processes, but will also suggest new applications.

Most of the literature deals with the liquid-solid interface. Probably for reasons of experimental procedure, the phenomena of adsorption at the liquid-liquid and liquid-vapour interfaces have seemed to form a different part of physical chemistry. I am sure that much will be gained if the three interfaces are considered together. Consequently, although most of this book deals with the liquid-solid interface (for the simple reason that this interface has received most attention), references to the other two are included.

The adsorption of electrolytes generally is dealt with elsewhere, in textbooks on electrochemistry, dyeing, detergency, ion-exchange, crystal growth, and classical colloid science. There is, inevitably, some overlap, and I have included a chapter to indicate where the two subjects are related.

To the best of my knowledge, a monograph on this subject has not appeared before in English, if in any other language. I have not, however, used this as a reason for unduly restraining the expression of my own views and prejudices. The author of a monograph of this kind is in one sense a historian, for the story which a historian writes is "really a series of answers to a set of questions".[1] The questions which I have tried to answer are my own selection, though I have also left some unanswered for the attention of the reader. If some answers which are desperately wanted have not been given, this may reflect my own inadequacy as a historian, or it may show that the men who make history have not yet thought about the questions. An obvious example is the paucity of studies of kinetics as compared with equilibrium. In trying to follow the far from uniform and, as it now seems, far from logical development of the subject, I have seen the making of a new set of questions, and hope that I have indicated the nature of some of them.

My thanks are due to all those from whom I have been able to learn, whether formally or in friendly discussion. I am grateful to the three professors of chemistry who, respectively, introduced me to adsorption from solution, encouraged me to pursue the subject in a small but growing depart-

[1] H. L. Short, *Hibbert Journal*, 1960, **58**, 217.

ment, and gave me opportunities to continue research and to write about it at a time when the calls of administration were particularly heavy. Especially am I grateful to the research students who have patiently shared with me the slow but absorbing process of learning a little more about the fascinating world of two dimensions.

hee-hee!

The University, Hull.   J. J. Kipling.
May, 1965.

CONTENTS

CHAPTER I

Introduction

TYPES OF INTERFACE

Adsorption from solution can take place at any of the three interfaces:

liquid–vapour
liquid–liquid
liquid–solid.

Of these the first two are in principle the simplest, because the adsorbed phase should be homogeneous if pure substances are used under clean conditions. By contrast, it is now recognized that most solids used as adsorbents are markedly heterogeneous, with poorly characterized surfaces. In practice, however, most attention has been given, and is still being given, to the liquid–solid interface. This interface must therefore form the basis of a general discussion of adsorption from solution at the present time.

In the past, little attempt has been made to compare adsorption from a given solution at the three types of interface. Two reasons for this may be suggested. The first is that different experimental methods of measuring adsorption are used (cf. Chapter 2). For the liquid–solid interface emphasis is centred on the change in concentration of the bulk liquid, whereas for the other two interfaces, the emphasis is on interfacial tension, a property of the interface itself. Secondly, adsorption at the liquid–solid interface is conventionally expressed in terms of unit weight of solid, whereas for the other two interfaces, it is expressed in terms of unit area of surface. This has militated against attempts at direct comparison, especially as the specific surface areas† of adsorbents used in early experiments were not usually known.

Recently, however, solids of well-defined specific surface area have become available and adsorption can now be expressed in terms of the same unit for all three interfaces. The unit of micromoles per square metre is then found to be convenient.[1] (This is equivalent to moles $\times 10^{-10}$ per sq. cm., which has sometimes been used for the liquid–vapour interface in the past.) Moreover, whereas the theory of adsorption at the liquid–solid interface has developed separately from that of adsorption at the other two interfaces, it may soon be possible, in simple cases, to apply a single theoretical treatment to all three interfaces. Some aspects of this development are considered in Chapter 14.

† The specific surface area is the surface area of unit weight of a solid.

ASPECTS OF ADSORPTION FROM SOLUTION

Most studies of adsorption from solution have been concerned with equilibrium conditions and predominantly with the *adsorption isotherm*. This term can have more than one meaning in this field. It is therefore important to differentiate the possible meanings and to distinguish them from the more familiar usage in adsorption of gases by solids, cf. especially Chapters 4 and 11.

In relatively few cases has adsorption been studied at different temperatures for a given system. When this has been done, it has usually been concluded that adsorptive forces weaken as the temperature rises. Such studies are likely to become more important as greater attention is paid to the thermodynamics of adsorption.

The kinetics of adsorption has also received little attention. In general, the extent of adsorption has been followed as a function of time solely for the purpose of determining when equilibrium has been reached. It is useful here to compare the results of kinetic studies on non-electrolytes with those made on ionic detergents. The latter are more numerous and have a number of relevant features.

The dynamic aspect of adsorption is also important when solid adsorbents are used in columns, as in a number of industrial processes and in chromatographic separations. The theoretical treatment of such procedures has dealt more extensively with the mathematics of flow through columns than with the nature of the adsorption processes as such.

The thermodynamic treatment of adsorption began effectively with the Gibbs equation, which has been an essential feature of the treatment of the liquid–vapour and the liquid–liquid interfaces. The treatment of the liquid–solid interface has, so far, been less coherent, but significant developments are now taking place.

TYPES OF ADSORPTION

In most of the systems under review, adsorption is due to the operation of physical forces only, and this has been an implicit assumption in the theoretical treatment of adsorption from solution. Solids, however, are known to adsorb gases chemically, even at low temperatures, and it has recently been found that chemisorption from solution can occur at room temperature. Only a few examples are yet known, but more may be expected to be discovered as more systems are investigated. Chemisorption may always be accompanied by physical adsorption onto the chemisorbed layer when a solution is involved.

This book is about the adsorption of non-electrolytes, but a number of useful points might be missed if too rigid a distinction between electrolytes and non-electrolytes were adopted. Some very interesting and technically important substances fall into an intermediate position as far as adsorption is concerned. Thus for electrolytes with large ions (e.g. some dyestuffs, detergents, and poly-electrolytes), adsorption may be governed largely by van der Waals forces because the ion is large in relation to the magnitude of its charge. A brief summary is given in Chapter 15 of some main features of similarity and of difference in the adsorption of non-electrolytes and of this kind of electrolyte.

In the adsorption of electrolytes by ionic solids, the possibility of ion-exchange must also be considered. Further, in any adsorption process involving weak electrolytes, it is important to consider whether the neutral molecule or the ion is adsorbed or even if both can be adsorbed.

CLASSIFICATION OF SYSTEMS

Almost all investigation of adsorption from solution has been carried out on two-component mixtures. This necessarily limits the strict application of theoretical concepts to practical use, in which multi-component systems are common. There is little point, however, in attempting a theoretical treatment of more complex systems until the behaviour of the simpler systems is adequately understood.

Analysis of multi-component systems presents considerable difficulty, as does the representation of results. Thus for a three-component system, a three-dimensional model would be required to represent adsorption equilibrium, the composition of the equilibrium liquid phase requiring a triangular diagram, and the extent of adsorption requiring surfaces in the third dimension.

Two-component systems can conveniently be classified as follows:

    (i) complete miscibility: two liquids
   (ii) partial miscibility:
      (a) solid in liquid
      (b) liquid in liquid
      (c) gas in liquid.

When a unifying theory of adsorption from solution has been fully developed, this classification may come to have less significance. At present, however, some of the differences in adsorption shown by the different systems seem to be important, though there are, of course, many similarities. This classification is applied, as far as possible, to the several interfaces discussed below.

DISCONTINUOUS ISOTHERMS

It is normally assumed that adsorption from solution, however expressed, is a continuous function of the concentration of the solution. There have been a few claims that discontinuous isotherms showing a large number of distinct breaks at short intervals of concentration have been observed in adsorption by solids.[2, 3] This parallels the observation of discontinuous isotherms in the adsorption of gases by solids.[4] In neither case has the original observation been followed up, and its significance will remain obscure until the experiments are repeated and either confirmed or refuted.

Discontinuous isotherms of this kind are not the same as "stepped" isotherms. The latter have been observed in adsorption both from the gas phase and the liquid phase (see Chapters 4 and 5). In the adsorption of gases they usually correspond to the formation of successive molecular layers of adsorbate on a few particularly homogeneous surfaces. In adsorption from solutions a variety of causes may be responsible.

REFERENCES

1. J. J. Kipling, "Proceedings of the Third International Congress of Surface Activity", Verlag der Universitätsdruckerei Mainz G.m.b.H., Cologne, 1960, Vol. II, p. 77.
2. R. Chaplin, *J. Phys. Chem.*, 1932, **36**, 909.
3. C. Ockrent, *Nature*, 1932, **130**, 206.
4. A. J. Allmand, L. J. Burrage, and R. Chaplin, *Trans. Faraday Soc.*, 1935, **28**, 218.

CHAPTER 2

Experimental Methods

GENERAL PROCEDURES

Adsorption at the Liquid–Solid Interface

Three main types of experiment are possible for measuring adsorption from liquids by solids. The essence of the first, which is by far the most frequently used, is to bring together known weights of solid and liquid, allow equilibrium to be established at a given temperature, and determine the resulting change of composition in the liquid.

The simplest effective procedure is to seal the solid and liquid (criteria for purity of which are discussed below) into clean, dry test-tubes. Loss of volatile components can be minimized by freezing the mixture (e.g. in a carbon dioxide slush or in liquid air) immediately before the tube is sealed. The tubes are then shaken in a constant temperature bath until equilibrium is reached. The time required for this stage varies considerably according to the type of system being investigated (see Chapter 13).

While adsorption is taking place, the liquid must be in constant motion and the solid must be continuously washed by the liquid. For many systems a reciprocating shaker is suitable, but a high frequency should be avoided, as this may lead to mechanical breakdown of the adsorbent, with consequent increase in specific surface area and adsorptive capacity.[1] Tumbling is often more effective, especially for dense solids. The rate of tumbling must be adjusted to allow the solid to fall completely through the liquid each half-cycle. A suitable tumbling apparatus is described by Giles.[2]

When finely divided powders are used as adsorbents, it is frequently necessary to centrifuge the system before the supernatant liquid is removed for analysis. This is best done before the tubes are opened if the liquid is at all volatile. Substantial changes of composition of some liquid mixtures can occur during the period of centrifuging if this is done with open tubes. Methods of analysing the solutions are described below.

If adsorption is carried out substantially above room temperature, it is useful to seal the solid and liquid unto a U-shaped tube having a constriction in the centre.[3] The solid and liquid are kept in one limb until equilibrium is established. The liquid is then run into the second limb, and both limbs are cooled in a freezing mixture. The second limb is sealed off at the constriction, and its contents are analysed after they have returned to room temperature. Separation at the temperature of the experiment is important.

If the system is allowed to cool down while separation is taking place, not only may the equilibrium change but, in cases of limited solubility, some of the solute may be precipitated.

An enclosed apparatus has been designed[4] for experiments to be carried out at temperatures above 100°c.

The second type of experiment is chromatographic and is based on the frontal analysis procedure of Tiselius. Its application to the investigation of adsorption has been described in detail by Claesson.[5] It is useful mainly for dilute solutions. In essence, the adsorbent is contained in a short column initially filled with the pure solvent. The solution is then run through the column and the concentration of the eluate is measured continuously; interferometric analysis (see below) is particularly valuable. The concentration is plotted as a function of the volume of eluate collected and, for a two-component solution, normally gives a curve of the form shown in Fig. 2.1. The volume collected up to point A is the volume of solution which

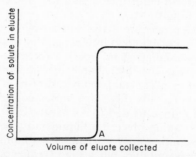

FIG. 2.1. Chromatographic method of measuring adsorption.

has been denuded of solute; allowance must be made for the volume of liquid held mechanically in the column containing the adsorbent. If, beyond the sharp rise, the concentration of the eluate which is collected remains constant, the solution and adsorbate in the column are in equilibrium. The adsorption isotherm can therefore be plotted from a series of such curves. The method can also be extended to deal with the simultaneous adsorption of two solutes.

In principle the Gibbs adsorption equation† can be used as the basis of a third method. The surface excess of component 2 adsorbed from a binary solution is given by:

$$\Gamma_2^{(1)} = -\frac{1}{RT}\frac{dS}{d\ln a_2}, \qquad (2.1)$$

where S is the interfacial tension, and a_2 the activity of component 2 in the liquid phase. No method for measuring S directly is known, but Bartell and

† The Gibbs equation is considered in greater detail in Chapter 11.

Benner[6] have shown that it can be replaced by the adhesion tension, A_{sl}, for $A_{sl} = S_{sa} - S_{sl}$, where S_{sa} and S_{sl} are the surface–air and surface–liquid interfacial tensions respectively. The former may be regarded as a constant; hence $dA_{sl} = -dS_{sl}$ and

$$\Gamma_2^{(1)} = \frac{1}{RT} \frac{dA_{sl}}{d \ln a_2}. \tag{2.2}$$

Bartell and Benner described a method of obtaining A_{sl}, but very few systems have been investigated in this way.

A few other methods have been used either on isolated occasions or for specific purposes. Thus Solov'eva[7] used the polarization of a zinc electrode to measure the extent and rate of adsorption of aliphatic alcohols on the electrode surface from aqueous solutions. Similarly measurement of hydrogen overvoltage at silver electrodes has been used to determine adsorption of n-heptylic and n-octanoic acids from aqueous solution.[8] The method is particularly useful if the adsorbent has only a small surface area.

For dilute solutions, surface potentials can be used to give an indirect measurement of adsorption by metals. A characteristic surface potential is observed when pure solvent is in contact with the metal. Differences between this and solution-metal potentials are attributed to rearrangement of dipoles at the surface when solute is adsorbed, and are considered to be directly proportional to the extent of adsorption of the solute.[9,10]

Adsorption at the Liquid–Vapour Interface

A few mechanical methods for direct determination of the composition of the adsorbed layer have been devised, but they have been little used. Those of McBain, in particular, are remarkable primarily for their experimental ingenuity. The direct methods also include the use of solutes labelled with radio-active isotopes. The commonest procedure is indirect, and is to deduce the extent of adsorption from the Gibbs equation (Chapter 11), for which the surface tension must be known as a function of the concentration of the bulk phase.

Direct Methods

(a) The Microtome Experiment

In the first of McBain's methods[11-14] a microtome, mounted on a trolley, was projected at high speed (35 ft./sec.) across the surface of a liquid, skimming off the surface layer to a depth of 0·05 mm., and directing it into a container held on the trolley. The composition of this liquid was compared with that of the original liquid interferometrically.

(b) Sweeping the Surface

In the second method[15] a mechanical barrier was pushed across the surface of a liquid, submerging the "adsorbed solute" and forcing it to dissolve in the underlying solution; a thin layer of this was swept by the barrier into one half of an interferometer cell, the other half of which contained a portion of the original liquid. (For this purpose the gas interferometer was used, so that the barrier could destroy a surface 72·5 cm. long.) The results were expressed in terms of sorption of the solute in the original surface.

(c) The Foam Method

A less well-known method was devised much earlier by Donnan and Barker.[16]   A stream of bubbles of regular size is passed, through baffles, up a column of the solution to be investigated. The rate is adjusted to allow adsorption at the liquid–air interface to be completed within a defined volume of liquid at the base of the column. The adsorbed material, released when the bubbles break at the top of the column, is not allowed to return to the base. After a suitable time, the liquid at the base of the column is analysed, and the deficiency in solute can be expressed as a corresponding surface excess at the total area of the bubbles formed in a given time. More recently this method has been used in studying the adsorption of surface-active agents at the liquid–vapour interface.[17]

(d) The Platinum Ring

This method[18] includes features of both the mechanical and radio-assay methods. Solutions are made from a solute labelled with a radio-active isotope. A portion of the surface film is removed by drawing a platinum ring through it. The film is then transferred to a weighing bottle and weighed. The quantity of solute which it contains is compared, by radio-active assay, with that present in the same weight of solution drawn from the interior of the liquid.

(e) Direct Surface Counting

The "direct" surface count method has been used extensively for investigating "surface-active" materials. The substance to be traced is labelled with an isotope emitting soft β-radiation, e.g. ^{3}H, ^{14}C, ^{35}S, ^{45}Ca, and the intensity of radiation is measured directly at the surface.[19-23] Counting techniques are generally simplified because the films are so thin that self-absorption of emitted radiation by the radio-active material is negligible. On the other hand, little radiation from the bulk solution can be measured as long as soft β-emitters are used. Method (d) is not limited in this way to isotopes emitting soft radiation.

(f) Gas–Liquid Chromatography

In gas–liquid chromatography, the gas retained by the liquid adsorbed on the column consists of that present in solution in the bulk of the liquid, and that adsorbed at the liquid–vapour interface. When thin films of liquid are used, the latter may be a significant fraction of the whole. If the total retention volume of the gas is V_R^0, then:

$$V_R^0 = kV_L + k_a A_L, \tag{2.3}$$

where V_L is the volume of liquid on the column, A_L the area of the liquid–vapour interface, k the partition coefficient, and k_a the adsorption coefficient.[24] Both A_L and V_L can be measured, and V_L can be varied systematically; hence k and k_a can be determined. The last two terms relate the concentration of vapour in the bulk liquid and adsorbed phases, respectively, to that in the vapour phase. Hence $\dfrac{k_a}{k}$ is a measure of the surface excess, and if calculated in the appropriate units, can be equated to Γ.

Figure 2.2 shows the result of determining the surface excess of a number of hydrocarbons in 1-chloronaphthalene by this method, the results being

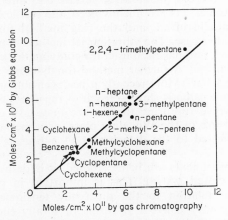

FIG. 2.2. Comparison of two methods of measuring surface excess.[24] (Reprinted with permission from *Analytical Chemistry*, copyright 1963 by the American Chemical Society.)

compared with those obtained from surface tension measurements by use of the Gibbs equation. It is important in this method that there should be no adsorption of the vapour at the interface between the liquid in the chromatographic column and its solid support.

Indirect Determination

This depends on measuring the surface tension; the surface excess is then calculated according to the Gibbs equation (Chapter 11). In general, activity data are also required. These can be obtained by various methods. Probably the most suitable for completely miscible liquids is to determine the partial pressures at the same temperature as the surface tensions. The "equilibrium still" gives the most accurate results;[25-28] a simpler apparatus, which gives results of sufficient accuracy for many systems, is described by Hovorka and Dreisbach.[29]

Standard methods of measuring surface and interfacial tensions are described by several authors.[30-35] The choice of method depends on the type of measurement being made. Thus some methods are not as suitable for solutions as for single liquids. There is also an important distinction between static and dynamic methods. The former are valuable for determining the surface tension of a system which has reached a steady condition. The dynamic methods are based on a surface which is continuously renewed. This may be an advantage in examining the properties and development of a freshly-formed surface, e.g. in kinetic studies, but a disadvantage in measuring the properties at equilibrium if adsorption occurs relatively slowly at the surface.

Harkins advocated the drop-weight method (for which he proposed important corrections) as the best method for surface and interfacial tensions of solutions if long time effects are not involved, the sessile bubble method if long time effects are involved, and the ring method for surface tension measurements in either case. The vibrating jet method is particularly useful as a dynamic method, and has been used for studying the kinetics of adsorption, e.g. by Hansen.[36, 37] A further method for dynamic studies is the use of the contracting jet.[38, 39] More recently it has been shown that for systems in which the surface tension is slow to reach or regain its equilibrium value, methods which involve rupturing the surface may give erroneous results. For this reason the use of the Wilhelmy plate, as a method based on true equilibrium, is preferred to the detachment of a ring as in the du Noüy method.[40]

When volatile components are being studied, it is desirable that they be confined within a closed system. This is important in most methods; a specific warning in relation to the du Noüy tensiometer is given by Hommelen.[41] Some very convenient forms of totally enclosed apparatus have been devised for the capillary rise method. This method has also been adapted for measurements on very small volumes of solution (0·01–0·03 ml.), with an accuracy of one part in a thousand.[42]

Adsorption at the Liquid–Liquid Interface

(a) The Emulsion Method

This is a direct method precisely analogous to the "foam" method used for the liquid–vapour interface. It has been used by Cockbain[43] for measuring adsorption of ionic detergents at the water–oil interface. It has also been used for investigating the adsorption of dyestuffs at interfaces between water and other liquids, the second liquid being allowed to rise, or fall (e.g. mercury), in the form of drops through a column of aqueous solution.[44, 45]

(b) Interfacial Tensions

The more usual method is to measure interfacial tensions and apply the Gibbs equation. Again, it is important to distinguish between methods designed for equilibrium conditions and those for dynamic conditions.

A recent description of the drop-weight method for measuring interfacial tensions is available.[46]

(c) *Adsorption from Mixed Vapours*

The relevant experiments here are those in which the vapour is saturated relative to the liquid mixture at the temperature of adsorption. The simplest arrangement is shown in Fig. 2.3. A stoppered tube is used to contain a known weight of liquid mixture of known composition, and a known weight

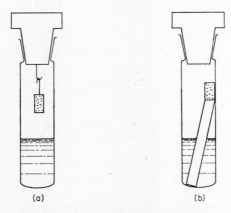

FIG. 2.3. Adsorption from mixed vapours.

of solid, contained in a perforated bucket, is held in the vapour space. The bucket may either be suspended, (a), or supported on a thick glass rod, (b). The bucket is weighed at intervals. When the weight has reached a constant value, the composition of the liquid is determined. Calculation of the amounts adsorbed is described in Chapter 4. Ideally, correction should be made for the weight of vapour present in the free space. In practice, it can be arranged so that this is so small that it can be neglected unless the extent of adsorption by the solid is very small.

Tryhorn and Wyatt[47] have shown that three stages occur in this type of experiment. This is illustrated in Fig. 2.4 for adsorption by charcoal from a mixture of acetone and benzene. In the first stage, acetone is adsorbed to a much greater extent than benzene. In the second stage (which presumably follows the complete coverage of the surface) no appreciable increase in total adsorption occurs, but the composition of the adsorbed phase changes. The third stage is regarded as capillary condensation and is effectively isothermal distillation of the bulk liquid into the pores of the adsorbent. For systems in which there is a sharp break between the second and third stages, the extent of primary adsorption of each component can be determined by weighing the adsorbent at the time when the liquid ceases to change

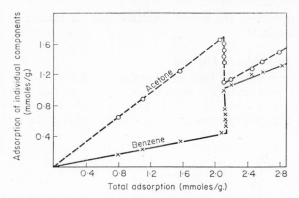

FIG. 2.4. Adsorption from mixed vapours of acetone and benzene: the initial liquid contained 55·2% of acetone by weight.[47] (Reproduced with permission from the *Transactions of the Faraday Society*.)

in composition. Other methods are more elaborate experimentally and have been devised primarily for the study of mixed vapours at less than saturation pressure.[48-50]

From their analysis of adsorption from vapour mixtures over a range of pressure, Cines and Ruehlen[51] plotted a relative distribution coefficient, K_R, as a function of pressure or surface coverage. K_R has a value of one when no preferential adsorption occurs. It is a coefficient of integral selectivity, as it is calculated from the composition of the complete adsorbed phase present at a given pressure. The differential selectivity can be obtained from the slope of the K_R versus pressure curve. When this is unity, primary adsorption has ceased.

METHODS OF ANALYSIS OF SOLUTIONS

In almost all studies of adsorption by solids from solution, it is necessary to measure the change in concentration of the solution which results from adsorption. In some cases a direct comparison is made of samples taken from the solution before and after adsorption. More frequently, separate analyses of the initial and final mixtures are made. The calculation of results is usually easier if the original mixtures are made up by weight and not by volume.

A variety of analytical methods has been used, mostly based on standard procedures. The following account is intended to indicate the wide range of methods which have been found useful, rather than to list every procedure or to describe all procedures in detail. Where further detail is required, textbooks of analysis (including those describing physical methods) or original papers should be consulted.

For dilute solutions, conventional titrimetric methods, or simple adaptations, are often useful. For completely miscible binary liquids, however, it would be necessary to measure a relatively small change in concentration (over most of the available range) as a difference between two large titres, and the measurement would be subject to a large experimental error. It is necessary, therefore, to rely on other properties for such systems, chosen so that the change in concentration can be measured more directly.

Refractometry

The use of refractive index very frequently meets the above requirements for completely miscible systems. It may also be useful for systems having a limited range of concentration if neither component possesses a functional group suitable for simple titration.

(a) The Pulfrich Refractometer

This instrument has a main scale, from which readings of refractive index can be taken, and a vernier scale, from which changes in refractive index can be read to the equivalent of the fifth place of decimals. A small difference in refractive index of two liquids can be read very accurately from the vernier scale without determining the actual refractive index of either. Thus an accurate setting of the main scale is made for the original mixture and the scale is clamped. The original mixture is then replaced by the final mixture, and a new setting is obtained by turning the vernier screw.

A calibration curve of refractive index against composition is made and from the slope of this curve at the point corresponding to the initial mixture, the change in composition can be determined and hence the composition of the final mixture. Occasionally, a very big change in composition is obtained, and the slope of the curve at the point corresponding to the final mixture is appreciably different from that for the initial mixture. In such a case, the composition of the final mixture can be found by successive approximations.

The change in composition of the liquid in a given experiment depends on (a) the difference in the refractive indexes of the two components, (b) the shape of the refractive index-concentration curve (e.g. whether it is approximately linear or deviates significantly from linearity), (c) the degree of preferential adsorption, (d) the weight ratio of liquid and solid used. When the change in refractive index is small, only the last factor can be adjusted to make a favourable measurement.

Results are accurate only if monochromatic light (e.g. from a sodium lamp) is used and if the prism and cell are maintained at a constant temperature by circulating water from a constant-temperature bath through the instrument. Even this may not be sufficient if the liquids are very volatile. Ioffe[52] has designed a cell for such systems in which the wall of the cell takes the form of a ground-glass cone. A thermostat head, with a corresponding socket, fits over the cell and seals it completely, except for a very small hole, through which liquid can be introduced into the cell from a hypodermic syringe. This resembles the cell designed by Miller[53] into which a liquid could be distilled directly without exposure to the atmosphere.

(b) Differential Refractometer

Greater sensitivity, corresponding to the sixth place of decimals in refractive index, is obtained from differential refractometers.[54–56]

(c) The Rayleigh Interferometer

This instrument, though slower to use than the refractometer, gives differences in refractive index to the seventh place of decimals. It is a differential instrument only, the reading obtained being a measure of the difference in optical path-length between two liquids being compared; this is obtained by a "null" method. The drum reading is converted into a difference in composition by means of a calibration curve. This is drawn by successive comparisons of a set of mixtures of accurately known composition, covering the relevant range of concentration. Values plotted as successive points are cumulative, so considerable care is required in constructing the calibration curve.

The use of the interferometer instead of the refractometer is governed partly by the difference in refractive index of the two components (factor (a) above), and partly by the combination of factors (c) and (d). Thus, even if there is a big difference in refractive index between the components, it may still be necessary to use the interferometer if the area of the interface is very small. In general, however, the use of the interferometer is restricted to systems with a small difference in refractive index, otherwise a large number of standard mixtures have to be made up for calibration.

For many liquid systems, the open cell normally supplied with the interferometer is inadequate, as differential volatilization alters the reading. A sealed cell must be used in such cases. A suitable cover was designed by Bartell,[57] but a specially made cell is now available[58] having a glass cover fused on to it, the cover having two filling holes which can be closed with ground-glass stoppers. An alternative design has been described by Grunwald and Berkowitz.[59]

A strict control of the temperature of the refractometer is necessary, but the interferometer is rather less sensitive to the precise temperature, provided that this does not fluctuate rapidly, or far from the standard temperature. This is because the temperature coefficients of refractivity of the two mixtures being compared are almost identical, as the difference in composition is small. A small thermostat for the cell is made, however; alternatively a wooden box to surround the instrument can be made which shields the cell effectively from draughts which might set up a temperature gradient. Further details of the instrument are given by Candler.[60]

Other Optical Methods

Simple colorimetry can be used in three ways: (a) directly for a coloured substance, (b) by converting a colourless substance to a coloured derivative, e.g. phenol is coupled with diazotized p-nitraniline and the colour compared with that given by standard solutions, (c) by causing the coloured substance to destroy the colour of a suitable reagent, e.g. fatty acids and cholesterol destroy the colour of warm sulphuric acid-dichromate mixtures.[61]

More frequently, spectrophotometric methods are used. The absorption of solutions is measured at a frequency giving the greatest sensitivity (i.e. maximum absorption). A frequency in the ultra-violet is generally used, but some determinations have been made at frequencies in the infra-red. Polystyrene in cyclohexane solutions has been estimated[62] by using the wavelength 260 mμ (3846 mm.$^{-1}$).

Polarimetry has been used to measure the adsorption *in situ* of decylamine from aqueous solutions on flat polished surfaces of metals.[63]

Polymers are sometimes estimated by measuring the turbidity produced by the addition to the solution of a liquid in which they are not soluble. The precipitation of the polymer in this way may be followed by rapid coagulation. To avoid this it may be necessary to use a weak precipitant, to obtain a stable, though often weak, turbidity.[64]

Titrimetric

The use of this method is sufficiently obvious, many standard procedures being available. As an example, adsorption of fatty acids has frequently been examined by titration with aqueous alkali, even if the fatty acid was originally dissolved in an organic solvent. The extraction of the acid from the solvent seems to cause no difficulty (especially if warm ethyl alcohol is added),[65] but the validity of the method should be checked by titration of a known sample for each case. To cover the range of concentration corresponding to a complete isotherm, it may be necessary to use the standard reagent in more than one concentration.

Conductimetric and electrometric titrations have been used as alternatives to those carried out with a coloured indicator.

Gravimetric

If an involatile solute is dissolved in a volatile solvent, analysis can be effected by evaporating off the solvent from a sample of known weight and weighing the residual solute. In the evaporation stage, a vacuum desiccator or a vacuum oven can be used with advantage. This method has been used, for example, in studying the adsorption of long-chain fatty acids from organic solvents such as benzene.[66] For such systems the required conditions are not perfectly met, as it is found that a small percentage of the solute is lost in the time required for complete removal of the solvent. In favourable cases, allowance can be made for this on the basis of calibration tests.

This method is often suitable for analysing solutions of high polymers. It is sometimes desirable to precipitate the polymer present in a known weight of solution and filter it off before drying it, e.g. polystyrene can be precipitated from solution in methyl ethyl ketone by running the known weight of solution into five times its volume of methanol.[67] Freeze-drying of polymers has been used.[68]

Densitometric

The use of a sensitive hydrometer to measure small changes in concentration has been advocated, but the method has not been widely used. It is likely to require greater volumes of the liquid to be examined than does either the refractometric or interferometric method.

Alternatively, the temperature of the solution under test can be varied until a float of known density just remains suspended in the solution. For this purpose a thermostat which can be controlled to $\pm 0.002°c$ is required.[69] Rather than adjust the temperature accurately to this point, it is suggested that the velocity with which the float moves should be determined at several accurately known temperatures and a graph of these values be used to determine the temperature at which the velocity is zero.

The Use of Radio-active Tracers

For adsorption at solid surfaces, this method can be used in three ways. First, it can be used for measuring changes in concentration, particularly when the solutions are very dilute. Thus it is valuable for determination of specific surface areas of solids (Chapter 17) especially when these are low.[70] It should also prove particularly valuable in examining adsorption from pairs of chemically similar, and hence usually physically similar liquids, for which the methods described above would be inadequate. Examples are near neighbours in homologous series, or isomers such as ethyl acetate and methyl propionate. It has also been used to measure the rate of exchange between adsorbed molecules and those in solution.[71]

Secondly, the method is increasingly being used to determine the actual composition of the adsorbed layer. The fundamental difficulty in the use of any method lies in knowing how to separate the adsorbate from the bulk liquid. For porous solids, physical separation is virtually impossible, as some of the bulk liquid is held mechanically in the pores. In the particular case of adsorption by plane surfaces (such as metals), however, the method is finding increasing use.

Smith and Allen[72] followed the adsorption of aliphatic acids on metals by labelling the carboxyl group with ^{14}C. After removal of the metal from the solution, excess liquid was allowed to drain away, and the radio-activity of the surface was measured. A number of small corrections was made, including one for the deposition of acid on the metal by evaporation of solvent during the draining process. Smith and McGill[73] later eliminated the "drainage" correction by removing excess solution with blotting paper. They claimed that this could be done reproducibly. The same method has been used for examining adsorption on other apparently plane surfaces such as glass, quartz, and mica.[74,75] This method has the advantage that it is of much higher sensitivity than most others for examining adsorption by solids of low specific surface area. Its limitation is that it can only be applied to plane solids.

A third use is in following competitive adsorption between two solutes of limited solubility. If the first contains a radio-active tracer, its adsorption in the presence of known amounts of the second can be determined. Thus the displacement of Aerosol OTN from the water–air interface by Aerosol OT has been followed in this way.[76]

Spreading Films

If the solute has the necessary polar group to enable it to be spread on an aqueous substrate to give a coherent film, the techniques of the Langmuir-Adam trough can be used. The area occupied by the solute (after evaporation of the solvent), at a standard pressure, is proportional to the weight present. It is particularly important, if this method is used, that the adsorbates be of high purity. The method has been successfully applied to analysis of solutions of long-chain fatty acids[77, 78] and of alcohols and phenols[79] in organic solvents such as benzene. Equal volumes of solution before and after adsorption were spread on the aqueous substrate. For reasonable sensitivity the adsorbent should have a specific surface area of about 0·5 sq. m./g.

Miscellaneous

Measurements of dielectric constant,[80] and cryoscopic measurements[81] have been used for dilute solutions. Surface tension measurements (see above) have been used particularly for aqueous solutions, as the solutes normally cause a rapid fall in the value (e.g. ref. 82).

The use of the critical solution temperature is said to be accurate and rapid.[83] It has been used, for example, to determine the content of butyl alcohol in benzene; the critical solution temperature observed when the benzene is added to aqueous acetic acid is very sensitive to the presence of dissolved butyl alcohol.

Very few of the above methods are adequate when adsorbents of low specific surface area are examined. Progress in examining a wider range of systems still depends on the development of new methods of analysis.

PURIFICATION OF MATERIALS

Adsorbates

In adsorption by solids from the liquid phase, substances present in low concentration are often adsorbed preferentially. The presence of impurities in the adsorbates may therefore have an effect on the measurements quite out of proportion to the quantity present. The purification of the adsorbates is thus of great importance. Reliable criteria of purity, sometimes with comments on preparation or purification, are given in an extremely useful compilation by Timmermans.[84] For acknowledged solvents, the data and comparative accounts of methods of purification given by Weissberger and co-authors are also very useful.[85]

For adsorption by solids a practical check of purity can be made as follows, and is valuable in avoiding spurious results. Suitable quantities of adsorbent and, separately, of each adsorbate are shaken together for the length of time adopted for the adsorption experiments. The resulting supernatant liquid is then examined by the method of analysis chosen for the particular experiment. There should be no change in the property measured greater than the appropriate experimental error.

Adsorbents

Solid adsorbents may need treatment of two kinds before use. Most are capable of adsorbing water vapour from the atmosphere and should therefore be dried. This is usually done by heating for about 2 hours at 110°–120°c. It is sometimes claimed that this temperature is not high enough for complete drying of all solids.[86] If a higher temperature is used, however, care must be taken that the solid is not altered by being heated, e.g. that sintering, change of crystalline form, or alteration of the nature of the surface (such as loss of surface complexes from carbons) does not take place. The temperature of drying must therefore be carefully chosen for each solid.

Some authors consider that solids should be "out-gassed" before use and then be introduced to the liquid in the absence of air. Others have claimed that such a procedure does not affect the extent of adsorption.[87] No systematic effect was found in adsorption by charcoal from mixtures of carbon tetrachloride and methanol.[88] The question is still debatable, but the situation should become clearer as further work of the kind described in Chapter 6 is carried out.

The second type of purification which may be necessary is illustrated by some experiments with charcoal,[89] summarized in Fig. 2.5. Most charcoals contain mineral matter,[90] some or all of which can be leached out by appropriate liquids. Thus charcoals made from vegetable matter may contain several percent of inorganic oxides, carbonates, etc., some of which can be extracted with water, ethanol, and even acetone;[91] a further amount can be extracted by weak organic acids, and most of the remainder by hydrofluoric acid. When such liquids are used as adsorbates, the analytical results may reflect changes due to leaching as well as those due to adsorption. This is clearly shown in Fig. 2.5. The treatment needed to overcome this difficulty is to extract the solid with a solvent at least as powerful as the adsorbates to be used. Soxhlet extraction is often very effective, and must be followed by complete removal of the solvent. Carbon blacks may contain organic matter ("adsorbed tars") which are soluble in hydrocarbons.[92] These can also be readily removed by Soxhlet extraction with benzene.

The procedure suggested above for testing the purity of the adsorbates also checks the purity of the adsorbent. A negative result shows that both adsorbate and adsorbent are satisfactory. A positive result does not always

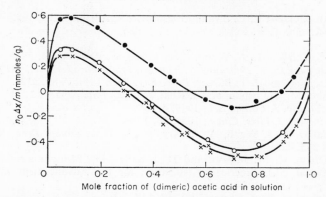

FIG. 2.5. Effect of inorganic salts present in coconutshell charcoal on adsorption from acetic acid–benzene mixtures: ●, original charcoal; ○, after extraction with water; ×, after extraction with acetic acid, followed by washing with water.[89] (Reprinted with permission from the *Journal of the Chemical Society.*)

show the source of the impurity. It may be present in the adsorbate, in the adsorbent, or in both.

Normally the isotherm obtained should be independent of the ratio of weights of solid and solution used. When this is not so, the presence of small quantities of impurity should be suspected. This is not the only cause. Thus, in adsorption from aqueous solutions by anodized aluminium, the effect was attributed to the slight dissolution of the alumina film in the water.[93]

It is worth noting that some potential adsorbents and adsorbates are incompatible. Activated alumina is dissolved by some solutions of acetic acid. It also accelerates the decomposition of t-butyl chloride, giving hydrochloric acid with which it reacts. With acetone it catalyses reactions which probably include polymerization.

REFERENCES

1. J. W. Galbraith, C. H. Giles, A. G. Halliday, A. S. A. Hassan, D. C. McAllister, N. Macaulay, and N. W. MacMillan, *J. Appl. Chem.*, 1958, **8**, 416.
2. A. Clunie and C. H. Giles, *Chem. and Ind.*, 1957, 481.
3. I. V. Smirnova, K. V. Topchieva, A. A. Kubasov, and V. V. Savchenko, *Proc. Acad. Sci. (U.S.S.R.)*, 1962, **147**, 836.
4. S. Y. Elovich, V. T. Avgul', and T. D. Semenovskaya, *Russ. J. Phys. Chem.*, 1963, **37**, 1037.
5. S. Claesson, *Arkiv Kemi. Min., Geol.*, 1946, **23** A, No. 1.
6. F. E. Bartell and F. C. Benner, *J. Phys. Chem.*, 1942, **46**, 847.
7. Z. A. Solov'eva, *Zhur. fiz. Khim.*, 1960, **34**, 537.
8. R. S. Hansen and B. H. Clampitt, *J. Phys. Chem.*, 1954, **58**, 908.
9. F. M. Fowkes, *J. Phys. Chem.*, 1960, **64**, 726.
10. D. A. Haydon, *Kolloid Z.*, 1961, **179**, 72.
11. J. W. McBain and C. W. Humphreys, *J. Phys. Chem.*, 1932, **36**, 300.

12. J. W. McBain and R. C. Swain, *Proc. Roy. Soc.*, 1936, A **154**, 608.
13. J. W. McBain, T. F. Ford, and G. F. Mills, *J. Amer. Chem. Soc.*, 1940, **62**, 1319.
14. J. W. McBain and L. A. Wood, *Proc. Roy. Soc.*, 1940, A **174**, 286.
15. J. W. McBain, G. F. Mills, and T. F. Ford, *Trans. Faraday Soc.*, 1940, **36**, 930.
16. F. G. Donnan and J. T. Barker, *Proc. Roy. Soc.*, 1911, A **85**, 557.
17. A. Wilson, M. B. Epstein, and J. Ross, *J. Colloid Sci.*, 1957, **12**, 345.
18. E. Hutchinson, *J. Colloid Sci.*, 1949, **4**, 599.
19. G. Aniansson, *J. Phys. Chem.*, 1951, **55**, 1286.
20. D. J. Salley, A. J. Weith, A. A. Argyle, and J. K. Dixon, *Proc. Roy. Soc.*, 1950, A **203**, 42.
21. J. K. Dixon, C. M. Judson, and D. J. Salley, in "Monomolecular Layers" (H. Sobotka, ed.), American Association for the Advancement of Science, Washington, D.C., 1954, p. 63.
22. J. E. Willard, *J. Phys. Chem.*, 1953, **57**, 129.
23. G. Nilsson, *J. Phys. Chem.*, 1957, **61**, 1135.
24. R. L. Martin, *Analyt. Chem.*, 1963, **35**, 116.
25. M. L. McGlashan, J. E. Prue, and I. E. J. Sainsbury, *Trans. Faraday Soc.*, 1954, **50**, 1284.
26. G. Scatchard, G. M. Kavanagh, and L. B. Ticknor, *J. Amer. Chem. Soc.*, 1952, **74**, 3715.
27. C. B. Kretschmer and R. Wiehe, *J. Amer. Chem. Soc.*, 1949, **71**, 1793.
28. C. B. Kretschmer, J. Nowakowska, and R. Wiehe, *J. Amer. Chem. Soc.*, 1948, **70**, 1785.
29. F. Hovorka and D. Dreisbach, *J. Amer. Chem. Soc.*, 1934, **56**, 1664.
30. N. K. Adam, "The Physics and Chemistry of Surfaces", Oxford University Press, London, 1941, 3rd edition.
31. W. D. Harkins in "Physical Methods of Organic Chemistry" (A. Weissberger, ed.), Interscience, New York and London, 1949, Vol. I, Part 1, 2nd edition.
32. W. D. Harkins, "The Physical Chemistry of Surface Films", Reinhold, New York, 1952.
33. J. J. Bikerman, "Surface Chemistry", Academic Press, New York, 1958, 2nd edition.
34. A. W. Adamson, "Physical Chemistry of Surfaces", Interscience, New York and London, 1960.
35. J. T. Davies and E. K. Rideal, "Interfacial Phenomena", Academic Press, New York and London, 1961.
36. R. S. Hansen, M. E. Purchase, T. C. Wallace, and R. W. Woody, *J. Phys. Chem.*, 1958, **62**, 210.
37. R. S. Hansen and T. C. Wallace, *J. Phys. Chem.*, 1959, **63**, 1085.
38. C. C. Addison and T. A. Elliott, *J. Chem. Soc.*, 1949, 2789.
39. F. H. Garner and P. Mina, *Trans. Faraday Soc.*, 1959, **55**, 1607.
40. J. F. Padday and D. R. Russell, *J. Colloid Sci.*, 1960, **15**, 503.
41. J. R. Hommelen, *J. Colloid Sci.*, 1959, **14**, 385.
42. J. Koefoed and J. V. Villadsen, *Acta Chem. Scand.*, 1958, **12**, 1124.
43. E. G. Cockbain, *Trans. Faraday Soc.*, 1954, **50**, 874.
44. C. W. Gibby and C. C. Addison, *J. Chem. Soc.*, 1936, 119, 1306.
45. C. W. Gibby and C. Argument, *J. Chem. Soc.*, 1940, 596.
46. F. Franks and D. J. G. Ives, *J. Chem. Soc.*, 1960, 741.

47. F. G. Tryhorn and W. F. Wyatt, *Trans. Faraday Soc.*, 1926, **22**, 139.
48. B. Lambert and D. H. P. Peel, *Proc. Roy. Soc.*, 1934, A **144**, 205.
49. W. K. Lewis, E. R. Gilliland, B. Chertow, and W. H. Hofmann, *J. Amer. Chem. Soc.*, 1950, **72**, 1153.
50. W. B. Innes and H. H. Rowley, *J. Phys. Chem.*, 1947, **51**, 1154.
51. M. R. Cines and F. N. Ruehlen, *J. Phys. Chem.*, 1953, **57**, 710.
52. B. V. Ioffe, *Zhur. fiz. Khim.*, 1960, **34**, 1133.
53. J. G. Miller, *J. Amer. Chem. Soc.*, 1934, **56**, 2360.
54. B. A. Brice and M. Halwer, *J. Opt. Soc. Amer.*, 1951, **41**, 1033.
55. S. J. Gill and D. B. Martin, *Analyt. Chem.*, 1963, **35**, 118.
56. A. G. Jones, "Analytical Chemistry, some new Techniques", Butterworths, London, 1959.
57. F. E. Bartell and C. K. Sloan, *J. Amer. Chem. Soc.*, 1929, **51**, 1637.
58. J. J. Kipling, *Hilger Journal*, 1957, **3**, 35.
59. E. Grunwald and B. J. Berkowitz, *Analyt. Chem.*, 1957, **29**, 124.
60. C. Candler, "Modern Interferometers", Hilger, London, 1951.
61. W. R. Bloor, *J. Biol. Chem.*, 1947, **170**, 671.
62. E. Treiber, G. Porod, W. Gierlinger, and J. Schurz, *Makromol. Chem.*, 1953, **9**, 241.
63. R. J. Ruch and L. S. Bartell, *J. Phys. Chem.*, 1960, **64**, 513.
64. E. Jenckel and B. Rumbach, *Z. Elektrochem.*, 1951, **55**, 612.
65. J. J. Kipling and E. H. M. Wright, *J. Chem. Soc.*, 1962, 855.
66. A. S. Russell and C. N. Cochran, *Ind. Eng. Chem.*, 1950, **42**, 1332.
67. J. F. Hobden and H. H. G. Jellinek, *J. Polymer Sci.*, 1953, **11**, 365.
68. F. M. Lewis and F. R. Mayo, *Ind. Eng. Chem., Analyt.*, 1945, **17**, 134.
69. H. Akamatu, *Bull. Chem. Soc. Japan*, 1942, **17**, 141.
70. M. C. Kordecki and M. B. Gandy, *J. Appl. Radiation and Isotopes*, 1961, **12**, 27.
71. R. N. Smith, C. F. Geiger, and C. Pierce, *J. Phys. Chem.*, 1953, **57**, 382.
72. H. A. Smith and K. A. Allen, *J. Phys. Chem.*, 1954, **58**, 449.
73. H. A. Smith and R. M. McGill, *J. Phys. Chem.*, 1957, **61**, 1025.
74. J. W. Shephard and J. P. Ryan, *J. Phys. Chem.*, 1959, **63**, 1729.
75. H. D. Cook and H. E. Ries, *J. Phys. Chem.*, 1959, **63**, 226.
76. C. M. Judson, A. A. Argyle, D. J. Salley, and J. K. Dixon, *J. Chem. Phys.*, 1950, **18**, 1302.
77. E. Hutchinson, *Trans. Faraday Soc.*, 1947, **43**, 439.
78. E. B. Greenhill, *Trans. Faraday Soc.*, 1949, **45**, 625.
79. D. J. Crisp, *J. Colloid Sci.*, 1956, **11**, 356.
80. L. Ebert and E. Waldschmidt, *Z. phys. Chem.*, 1931, Bodenstein Festband, 101.
81. H. L. Richardson and P. W. Robertson, *J. Chem. Soc.*, 1925, **127**, 553.
82. W. E. Garner, D. McKie, and B. C. J. G. Knight, *J. Phys. Chem.*, 1927, **31**, 641.
83. D. C. Jones, *J. Chem. Soc.*, 1923, **123**, 1384.
84. J. Timmermans, "Physico-Chemical Constants of Pure Organic Liquids", Elsevier, London, 1950.
85. A. Weissberger, E. S. Proskauer, J. A. Riddick, and E. E. Toops, "Organic Solvents", Interscience, New York and London, 1955.
86. P. Berthier, L. Kerlan, and C. Courty, *Compt. rend.*, 1958, **246**, 1851, cf. J. H. de Boer, "The Dynamical Character of Adsorption", Clarendon Press, Oxford, 1953, p. 42.

87. O. C. M. Davis, *J. Chem. Soc.*, 1907, **91**, 1666.
88. W. B. Innes and H. H. Rowley, *J. Phys. Chem.*, 1947, **51**, 1172.
89. A. Blackburn and J. J. Kipling, *J. Chem. Soc.*, 1955, 4103.
90. J. J. Kipling, *Quart. Rev.*, 1956, **10**, 1.
91. A. Blackburn and J. J. Kipling, *J. Chem. Soc.*, 1954, 3819.
92. G. Kraus and J. Dugone, *Ind. Eng. Chem.*, 1955, **47**, 1809.
93. C. H. Giles, H. V. Mehta, S. M. K. Rahman, and C. E. Stewart, *J. Appl. Chem.*, 1959, **9**, 457.

CHAPTER 3

Adsorption at the Liquid–Solid Interface

HISTORICAL

More is known about this interface than about the other two liquid interfaces because purification of liquids (drinking water, wine, oils) by solid adsorbents, especially charcoal, has been familiar for centuries. With the expansion of chemical industry over the last century, the range of substances purified by adsorption has increased enormously; a large-scale use developed quite early in sugar refining.[1]

Adsorption at this interface may, indeed, have been important as long ago as the time at which living organisms first developed from inanimate matter. Simple molecules, such as amino-acids, could have been formed in the seas by photochemical reactions. The solutions would, however, have been dilute, and a higher concentration would have been required before polymerization could have occurred to give the high-molecular-weight substances characteristic of living systems. Bernal[2] has suggested that adsorption on clay minerals could have effected this concentration and also provided a substrate on which the polymerization process could have occurred.

The term adsorption was generally used to describe the concentrating of a particular component at an interface relative to an adjacent solution or other bulk phase. As will be seen later, this concept needs more precise formulation. It formed the basis, however, for the early experiments on adsorption from solution.

The fundamental investigation of adsorption proceeded only slowly, in spite of its technological importance. It was overtaken by the study of adsorption from the vapour phase, partly as a result of the stimulus afforded by two world wars which suggested subjects for academic investigation as well as requiring the solution of practical problems. Although the development in this related field will ultimately prove to be valuable to the study of adsorption from solution, it has had at least one major adverse effect. Study of the adsorption of gases has usually been confined to a single component and this has helped to strengthen the existing concept of adsorption as applying normally to only one component, whether this be a vapour or a solute present in dilute solution.

This concept has serious limitations, especially when concentrated solutions are considered, because solutions necessarily consist of at least two components. A more useful concept of adsorption is now developing which

23

can be applied equally to dilute and to concentrated solutions. It can also be used more successfully than the more limited concept to relate the phenomena of adsorption which occur at different interfaces. (Adsorption at the liquid–solid interface now seems much more similar to adsorption at the liquid–vapour interface for a two-component system than to adsorption of a single vapour by a solid. A close connection with adsorption from saturated *mixed* vapours is discussed below.

EARLY EXPERIMENTS

Some of the early experiments with solutions are summarized by Freundlich.[3] A variety of substances, including phenol, succinic acid, the simpler aliphatic acids, and bromine were adsorbed by charcoal from dilute aqueous solutions. The importance of aqueous solutions is sufficient to explain their being

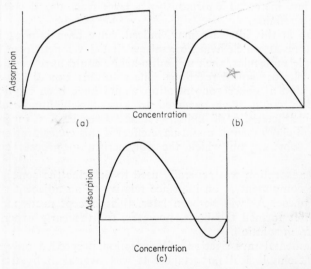

FIG. 3.1. Types of adsorption isotherm.[14] (Reprinted from *Quarterly Reviews* with the permission of The Chemical Society and the author.)

chosen for investigation, but in the light of later work they appear to be far from typical and often highly complex. Measurements were made mainly of adsorption isotherms, which, for these dilute solutions, had the shape shown in Fig. 3.1(a). These isotherms could be fitted by the equation:

$$\frac{x}{m} = \alpha . c^{1/n} \tag{3.1}$$

where x is the weight of adsorbate taken up by a weight m of solid, and c is the concentration of the solution at equilibrium; α and n are constants, the form $1/n$ being used to emphasize that c is raised to a power less than unity.

Equation (3.1) is usually known as the Freundlich equation, mainly because Freundlich used it extensively, but he was not its originator. Equations of the same form had earlier been used by Küster[4] to describe the partition of iodine between a solution in aqueous potassium iodide and starch, by Boedecker[5] to describe adsorption by soil, and by others.[6] Equation (3.1), with concentration replaced by pressure, was also used by Freundlich and others to describe the isotherm for adsorption of gases by solids. At that time it had no theoretical foundation, the lack of which hampered the development of a theory for either kind of adsorption. A justification for its application to adsorption from solution was subsequently given by Henry (see Chapter 11), but it has generally been overlooked that this is based on an approximation which is valid only for dilute solutions.

The Freundlich equation proved to be applicable to the adsorption of gases over only limited ranges of pressure. It was replaced by the Langmuir equation and others which had a theoretical basis in the kinetic theory of gases. Although the Langmuir equation has met with some success in its application to adsorption of gases, it is clear that neither it nor the Freundlich equation is of the right mathematical form to describe isotherms of the shape shown in Fig. 3.1(b) and (c). Neither provides for a maximum in the curve.

Several attempts have been made to derive a general equation for the adsorption isotherm. The early attempts are summarized by Swan and Urquhart.[7] As yet, no simple equation has been put forward which can be applied universally.† This undoubtedly arises from the wide range of complexity found in solutions. The most promising developments, as yet applicable only to relatively simple cases, are described in Chapter 14.

Many equations have proved inadequate because the role of the solvent has been neglected. Advances in treating the liquid–solid interface have depended on the recognition that the liquid phase contains at least two components. The nature of the adsorption isotherm has had to be reconsidered in the light of this.

THE NATURE OF THE ADSORPTION ISOTHERM

In the early experiments, charcoal was the most widely used adsorbent. Adsorption on charcoal gave isotherms of the form shown in Fig. 3.1(a), both for solutions of solids (e.g. benzoic acid in water, or stearic acid in benzene) and for solutions of liquids (e.g. acetic acid in water), between which no distinction seems to have been drawn at the time. The Freundlich equation was applied indiscriminately to the isotherms, all of which were

---

† The Gibbs equation does, of course, apply to adsorption at all interfaces, but involves a further term (interfacial tension) and thus does not describe adsorption solely in terms of the concentration of solution.

limited to adsorption from dilute solutions. Whereas this limit was imposed by the solubility of the solids, it was not imposed on the investigation of all solutions of liquids, yet it took many years for concentrated solutions to be explored, and many more for the results of these explorations to be widely appreciated.

Acetic acid and water are miscible in all proportions at room temperature. The isotherm for adsorption on charcoal, when plotted for the whole range of concentration, has the form shown in Fig. 3.1(b). As this passes through a maximum, it clearly cannot be fitted by a Freundlich equation, nor by any modification of the Langmuir equation. In adsorption from other mixtures of completely miscible liquids (e.g. ethyl alcohol and benzene), the isotherm may have the S-shape shown in Fig.3.1(c), i.e. it passes through both a maximum and a minimum and has a negative branch.

These results seem strange only because adsorption by solids from solution has habitually been treated as though it were analogous with adsorption of single gases. The data which are plotted as isotherms, however, are different in principle. The isotherm for adsorption of a *single* gas by a solid represents directly the quantity (weight, or volume under standard conditions) of gas adsorbed by unit weight of the solid. The experimental measurement in adsorption from solution is the *change in concentration* of the solution which results from adsorption. The fact that a change in *concentration* is measured emphasizes that there are at least two components in the solution.

It was formerly assumed that the change in concentration was a measure of the extent to which one component (the "solute") had been adsorbed. The extent of adsorption was given by multiplying the change in concentration by the weight of solution used. It was tacitly assumed that the second component (the "solvent") was not involved in the adsorption process, and could be regarded merely as "a 'space' in which the solute has play".[8] This concept may prove to be approximately valid for systems in which the "solute" has very limited solubility in the "solvent". For systems in which the two components are completely miscible, however (e.g. ethyl alcohol and benzene), neither can be regarded as solvent or as solute over the whole range of concentration. It thus becomes important to recognize that each component of the mixture may be adsorbed.

This cardinal feature of adsorption from solution seems first to have been clearly understood by Williams in 1913, and was later emphasized by Heymann and Boye.[9] It has, however, even now not been recognized by all text-books. Perhaps this demonstrates how important a place aqueous solutions, and particularly dilute aqueous solutions, have in the teaching of chemistry, and how difficult it is for chemists to regard water in a mixture as anything other than the solvent.

It is now possible to understand, qualitatively, the significance of isotherms of the type shown in Fig. 3.1(c). Over the first part of the concentration range, one component is adsorbed preferentially with respect to the first. This means that, at equilibrium, it is present in the adsorbed layer in greater proportion than in the bulk liquid. "Negative" adsorption of component 1 thus means preferential adsorption of component 2. For completely miscible liquids, the isotherm must fall to zero at each end of the concentration range, as no change in composition at these points (i.e. of the pure liquids) is possible. This point was emphasized by Patrick and Jones,[10] but has often been neglected.

Definitions of Adsorption

It is clearly important to recognize two concepts of adsorption as applied to mixtures. The first is given by the experimental measurements just described and is now referred to as *preferential* or selective adsorption.[11] (At one time, the term *apparent* adsorption was used.) This corresponds to the term *surface excess* which has been used to describe the same situation at the liquid–vapour interface (cf. Chapter 11). It is a measure of the extent to which the bulk liquid is impoverished with respect to one component, because the surface layer is correspondingly enriched. The second is *absolute* adsorption or adsorption of an *individual* component. This refers to the actual quantity of that component present in the adsorbed phase as opposed to its excess relative to the bulk liquid. It is a *surface concentration*.

Some consequences of the use of these two concepts have been examined by Baylé and Klinkenberg.[12] They emphasize that preferential adsorption is directly related to experimental measurements and can therefore be expressed directly and unambiguously. Their claim, that isotherms based on preferential adsorption should normally be used in the practical application of adsorption, can, to this extent, be supported. Attempts to determine absolute adsorption, however, have led to a much better understanding of the phenomena of adsorption and especially of the meaning of preferential adsorption. These attempts are therefore considered in detail in the following chapters, for both concentrated and dilute solutions.

The Composite Isotherm

Preferential or selective adsorption is represented by what was formerly called simply "the adsorption isotherm", but which must now be renamed to avoid ambiguity. A suggestion that it should be designated "the isotherm of concentration change" was an attempt to focus attention on the actual experimental measurements which it summarizes.[13] Its description as a "composite isotherm"[14] is, however, coming to be used most frequently. This emphasizes that it is the result obtained by combining the "true" (or individual) isotherms for adsorption of each component.

The significance of the composite isotherm is shown by deriving an equation to relate the preferential adsorption from a two-component mixture to the actual adsorption of each component. This derivation is not based on any supposed mechanism of adsorption except that each component of the liquid mixture may be adsorbed at the interface.

When a weight m of solid is brought into contact with n_0 moles of liquid, the mole fraction of the liquid decreases† by Δx with respect to component 1. This change in concentration is brought about by the transfer of n_1^s moles of component 1 and n_2^s moles of component 2 onto the surface of unit weight of solid. At equilibrium, there remain in the liquid phase n_1 and n_2 moles, respectively, of the two components, giving a mole fraction, x, with respect to component 1, the initial mole fraction having been x_0.

Then

$$n_0 = n_1 + n_2 + n_1^s m + n_2^s m,$$

and

$$x_0 = \frac{n_1 + n_1^s m}{n_0}, \quad x = \frac{n_1}{n_1 + n_2} \quad \text{and} \quad 1 - x = \frac{n_2}{n_1 + n_2}.$$

$$\therefore \quad \Delta x (= x_0 - x)$$

$$= \frac{n_1 + n_1^s m}{n_1 + n_2 + n_1^s m + n_2^s m} - \frac{n_1}{n_1 + n_2}$$

$$= \frac{n_1^2 + n_1 n_2 + n_1 n_1^s m + n_2 n_1^s m - n_1^2 - n_1 n_2 - n_1 n_1^s m - n_1 n_2^s m}{(n_1 + n_2)(n_1 + n_2 + n_1^s + n_2^s)}$$

$$= \frac{n_2 n_1^s m - n_1 n_2^s m}{(n_1 + n_2) n_0}$$

$$\therefore \quad \boxed{\frac{n_0 \Delta x}{m} = n_1^s (1 - x) - n_2^s x} \tag{3.2}$$

or

$$\boxed{\frac{n_0 \Delta x}{m} = n_1^s x_2 - n_2^s x_1,} \tag{3.3}$$

where x_1 and x_2 refer to the mole fractions of components 1 and 2, respectively, in the liquid phase.

The function $\dfrac{n_0 \Delta x}{m}$ is (when moles and mole fractions are used) what has actually been plotted as "adsorption" to give what is properly the composite isotherm. The use of mole fraction units is recommended whenever it is practicable. It is also appropriate because it is the unit adopted in the study of liquid mixtures (e.g. ref. 16).

The composite isotherm is sometimes expressed in terms of $(x_1^s - x_1)$, where x_1^s is the mole fraction of component 1 in the adsorbed layer. This form emphasizes the preferential adsorption of one component. It can readily be shown that

$$x_1^s - x_1 = \frac{1}{n^s} \frac{n_0 \Delta x}{m}, \tag{3.4}$$

† If the use of the operator Δ is taken to imply an increase, rather than a change in x, then $-\Delta x$ should be used in this context, as has been done by Elton,[15]

where n^s is the total number of molecules in the adsorbed layer on unit weight of solid. Hence

$$n^s(x_1^s - x_1) = n_1^s(1 - x_1) - n_2^s x_1. \tag{3.5}$$

An equation analogous to (3.2) can be derived in terms of weights and weight fractions:

$$\frac{w_0\,\Delta c}{m} = w_1^s(1-c) - w_2^s c, \tag{3.6}$$

where w_0 is the initial weight of liquid brought in contact with a weight m of adsorbent; the equilibrium *weight fraction* of the solution is c, and w_1^s and w_2^s are the weights of the two components, respectively, adsorbed by unit weight of adsorbent. Equation (3.6) is particularly useful when the molecular weight of one component is not known. It is implicit in Williams' treatment of adsorption[8] as it can be obtained by combining two of his equations.

(The use of concentrations expressed in moles per litre is not recommended. It is sometimes convenient for dilute aqueous solutions, but as density normally changes with concentration, the conversion of the adsorption measured in these units to either of the other units is cumbrous. It is more unsatisfactory for dilute solutions in organic solvents than for dilute aqueous solutions, because a given molarity corresponds to a much higher mole fraction for the former than for the latter. A number of treatments of adsorption have proved to be approximate because these points have not been appreciated.)

It is sometimes useful to know the connection between

$$\frac{n_0\,\Delta x}{m} \quad \text{and} \quad \frac{w_0\,\Delta c}{m}.$$

It is:

$$\frac{n_0\,\Delta x}{m} = \frac{w_0\,\Delta c}{m} \cdot \left(\frac{x}{M_2} + \frac{1-x}{M_1}\right)$$

$$= \frac{w_0\,\Delta c}{m} \cdot \left(\frac{1}{M_2 c + M_1(1-c)}\right), \tag{3.7}$$

where M_1 and M_2 are the molecular weights of components 1 and 2 respectively.

Equations (3.2) and (3.6) are exactly applicable to adsorption from all solutions, of whatever concentrations. Their use is considered in the following chapters, together with an indication of approximations which may be made for dilute solutions. The close connection between "preferential adsorption" at the liquid–solid interface ($n_0\Delta x/m$ or $w_0\Delta c/m$) and "surface excess" at the liquid–liquid or liquid–vapour interface, is considered in Chapter 11.

2*

Individual Isotherms

(a) Synthesis to give Composite Isotherms

The relationship between individual isotherms and the composite isotherm was shown in principle by Ostwald and de Izaguirre.[17] By postulating different forms for the individual isotherms, they calculated what form the resultant composite isotherm must have. Six examples are shown in Fig. 3.2.

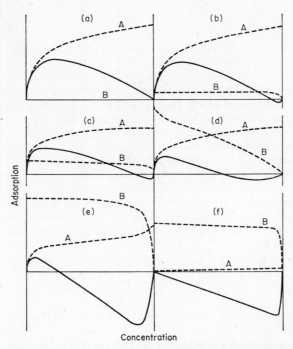

FIG. 3.2. Individual and composite isotherms.[14] Full line: composite isotherm. Broken lines: A, "solute"; B, "solvent". (Reprinted from *Quarterly Reviews* with the permission of The Chemical Society and the author.)

These figures are useful in illustrating that "negative" adsorption is only possible if the solvent is adsorbed (though adsorption of the solvent does not necessarily result in "negative" adsorption of the solute). They also show that a maximum in the composite isotherm can occur even without adsorption of solvent. Equation (3.2) then becomes:

$$\frac{n_0 \Delta x}{m} = n_1^s (1-x) \qquad (3.8)$$

and the decreasing value of $(1-x)$ results in a fall in the value of $n_0 \Delta x / m$, linearly with x.

(b) Analysis of Composite Isotherms

In practice, the main interest lies in resolving the composite isotherm into individual isotherms. Because equation (3.2) contains two unknowns, this cannot be done exactly without supplementary information relating to n_1^s or n_2^s, or both. The procedure used varies according to the type of system.

For very dilute solutions, an approximation is often possible. It follows from equation (3.2) that for small values of x (e.g. $x \leqslant 0 \cdot 01$)

$$n_1^s \sim \frac{n_0 \Delta x}{m}. \tag{3.9}$$

In such cases, even if n_2^s is considerable, the term $n_2^s x$ can be neglected, to a first approximation. Similarly, the term $n_1^s(1-x)$ approximates to n_1^s because $(1-x)$ is nearly equal to one. This situation often applies to solutions of solids in liquids because the solubility limit is low in terms of mole fractions. If the solid is very soluble in the liquid, however, the approximation is a bad one.

Although n_1^s can be obtained in this way, the corresponding value of n_2^s is not obtained, and for this another method is needed. A further method is, in any case, required to determine n_1^s and n_2^s for concentrated solutions. As this is immediately required for a discussion of completely miscible liquids, it is described in Chapter 4. Application to dilute solutions is described in Chapters 5, 6, and 7.

REFERENCES

1. V. R. Deitz, "Bibliography of Solid Adsorbents", National Bureau of Standards, Washington, D.C., 1944.
2. J. D. Bernal, "The Physical Basis of Life", Routledge and Kegan Paul, London, 1951.
3. H. Freundlich, "Colloid and Capillary Chemistry", Methuen, London, 1926.
4. F. W. Küster, *Annalen*, 1894, **283**, 360.
5. C. Boedecker, *J. Landwirtschaft*, 1859, **7**, 48.
6. J. W. McBain, "The Sorption of Gases and Vapours by Solids", Routledge, London, 1932, p. 4.
7. E. A. Swan and A. R. Urquhart, *J. Phys. Chem.*, 1927, **31**, 251.
8. A. M. Williams, *Med. K. Vetenskapsakad. Nobelinst.*, 1913, **2**, No. 27.
9. E. Heymann and E. Boye, *Kolloid Z.*, 1933, **63**, 154.
10. W. A. Patrick and D. C. Jones, *J. Phys. Chem.*, 1925, **29**, 1.
11. M. R. A. Rao, *J. Indian Chem. Soc.*, 1935, **12**, 345.
12. G. E. Baylé and A. Klinkenberg, *Rec. Trav. Chim.*, 1957, **76**, 593.
13. J. J. Kipling and D. A. Tester, *J. Chem. Soc.*, 1952, 4123.
14. J. J. Kipling, *Quart. Rev.*, 1951, **5**, 60.
15. G. A. H. Elton, *J. Chem. Soc.*, 1951, 2958.
16. J. H. Hildebrand and R. L. Scott, "Solubility of Non-Electrolytes", Reinhold, New York, 1950, 3rd edition.
17. W. Ostwald and R. de Izaguirre, *Kolloid Z.*, 1922, **30**, 279.

CHAPTER 4

Adsorption from Completely Miscible Liquids

Nearly all of the investigations on this type of system have been made with binary mixtures, and the treatment in this section refers to such mixtures.

THE INDIVIDUAL ISOTHERM

As equation (3.2) contains two unknowns, n_1^s and n_2^s, a second equation is needed before either term can be evaluated.[1] The ideal way of determining n_1^s and n_2^s would be to analyse the adsorbed layer, but no method has been devised for separating this from the bulk liquid in contact with it.[2]

Two ways of providing a second equation have been put forward. The first involves an indirect experimental measurement. The second involves a theoretical consideration of the nature of the adsorbed layer. Different assumptions can be made, and on the validity assigned to each depends the significance of the ensuing results.

The Experimental Approach

Williams suggested the first experimental approach. If, in a closed vessel, the solid is suspended in the vapour above (and thus in equilibrium with) the liquid mixture (Fig. 2.3) it is possible to follow the change in concentration of the liquid by the normal methods and to measure the total amount of adsorbed material by weighing. The necessary second equation is then:

$$w = w_1^s + w_2^s \tag{4.1}$$

or
$$w = n_1^s M_1 + n_2^s M_2, \tag{4.2}$$

where w is the observed weight, and M_1 and M_2 are the molecular weights of the two components respectively.

It can generally be assumed that the equilibrium at the solid–adsorbate interface established in this way is the same as would be established by immersing the solid in the liquid. In principle, there is one difference. In adsorption from mixed vapours, an extra interface—between the adsorbate and the vapour—is introduced, which does not exist in adsorption directly from the liquid phase. Adsorption at this interface is generally assumed to be negligible, but this assumption should be examined for every system to which the method is applied.

32

Williams attempted to apply this method to the adsorption by charcoal from mixtures of acetic acid and water. The choice of this liquid system was a natural one as it had been examined by a number of earlier investigators. It was nevertheless unfortunate in two respects. First, equilibrium was established so slowly that some of the mixtures were attacked by fungus before the final readings were taken; the experiment was therefore abandoned in an incomplete state. Secondly, the liquid system is one of great complexity from the point of interaction between the components,[3,4] and progress in the understanding of this type of adsorption came first from the examination of simpler systems. Williams did not plot his incomplete results in the form of individual isotherms, but this was done by Gustafson.[5] These are reproduced in Fig. 4.1 for comparison with later results for this system (Figs. 4.3 and 10.12).

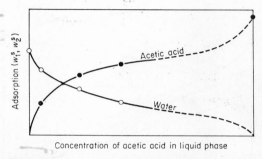

FIG. 4.1. Individual isotherms for adsorption by charcoal from mixtures of acetic acid and water. (Reprinted from *Quarterly Reviews* with the permission of The Chemical Society and the author.)

In later investigations, the method has been used much more successfully, especially in the adsorption by charcoal from binary mixtures,[6–9] including some aqueous mixtures.[10] The rate at which equilibrium is established depends mainly on the vapour pressures of the two components. It is quite rapid if the individual components have saturated vapour pressures of over 40 mm. Hg, but it may take several days or even weeks for equilibrium to be established if they are only a few millimetres. The attainment of equilibrium may be a stepwise process (see Chapter 2).

If the adsorbent has wide pores, two processes may occur in adsorption from the mixed vapours. Immediately adjacent to the surface, primary adsorption occurs, giving an adsorbate of composition different from that of the liquid phase. The remainder of the pore space could then be filled by a process of capillary condensation of the mixed vapours. This second process would give rise to a condensate having the same composition as the bulk liquid at equilibrium. The condensate is equivalent to bulk liquid which penetrates into the pores when adsorption takes place from the liquid phase;

it is held only mechanically in the pores, but it is nevertheless difficult to separate this extension of the bulk liquid from the material held by primary adsorption.

Jones and Outridge[11] measured adsorption by silica gel from a mixture of n-butyl alcohol and benzene vapours. They attempted to calculate the extent of primary adsorption by assuming that each individual isotherm was governed by a Freundlich equation (see Chapter 3), the constants being compatible with the form of the composite isotherm. The individual iso-therms obtained in this way (Fig. 4.2) fell below the curves (calculated from direct weighing) for the total uptake of each component. The difference was

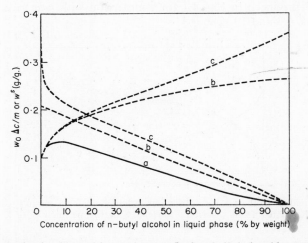

FIG. 4.2. Adsorption by silica gel from mixtures of n-butyl alcohol and benzene.[11] (a) Composite isotherm; (b) calculated individual isotherms; (c) total uptake of individual components. (Reprinted from the *Journal of the Chemical Society* with the permission of The Chemical Society and the author.)

ascribed to "capillary adsorption". (This phrase, although appropriate to adsorption from the vapour phase, has little meaning when used in descriptions of adsorption from the liquid phase.)

While it is necessary to distinguish between primary adsorption and either capillary condensation from the vapour phase or mechanical penetration of the pores by the bulk liquid phase, it is now doubtful whether the Freundlich equation gives a satisfactory quantitative description of primary adsorption. The difficulty in making this distinction by any other means is a limitation of the method as applied to porous solids with wide pores. For non-porous solids there is a similar difficulty in distinguishing between primary adsorption and subsequent multilayer condensation; in this case a close approximation is probably given by Cines and Ruehlen's method (Chapter 2).

A suggestion was made[12] that the difficulty caused by capillary condensation of the vapour could be eliminated if the adsorbent were held at a much higher temperature than the bulb containing the liquid from which the vapour was derived. This is not satisfactory, however, because the vapour is not a saturated mixed vapour at the temperature at which adsorption takes place, and the process is therefore not analogous to adsorption from the mixed liquid at either temperature.

The Theoretical Approach

(a) The Use of Equations Originally Applied to Adsorption of Gases

Several attempts have been made to combine equation (3.2) or (3.6) with an equation expressing the individual adsorption of each component as a function of its concentration in the bulk liquid phase. Thus Ostwald and de Izaguirre[13] assumed that component 1 was adsorbed according to a Freundlich equation:

$$w_1^s = k_1 c^\alpha. \tag{4.3}$$

They considered five theoretical cases, in which the second component was (i) not adsorbed at all, (ii) adsorbed according to a Freundlich equation:

$$w_2^s = k_2 (1-c)^\beta, \tag{4.4}$$

(iii) not adsorbed directly, but in combination with component 1 (this was expressed as adsorption of a solvated solute, the solvent not being otherwise adsorbed), (iv) adsorbed both directly and in combination with component 1. In the fifth case the "solvent" was adsorbed and the "solute" was not adsorbed; this gives the same type of result as in case (i).

The number of experimental data relating to the complete concentration range was, at that time, very small. The few available belonged to case (ii), for which the following equation had been derived:

$$\frac{w_0 \Delta c}{m} = k_1 c^\alpha (1-c) - k_2 (1-c)^\beta c. \tag{4.5}$$

This equation was found to fit very well the composite isotherms which had been obtained earlier for the adsorption by charcoal from acetic acid–water mixtures[14] and from ethyl alcohol–phenol mixtures;[15] these are U-shaped and S-shaped isotherms respectively. The corresponding individual isotherms for the former system are shown in Fig. 4.3, and may be compared with those given in Fig. 10.12 which were calculated on a different basis from similar data.

When they examined adsorption by charcoal from mixtures of acetic acid and ethylene dibromide, Ostwald and de Izaguirre obtained an S-shaped composite isotherm. As this was not readily fitted by an equation having the form of (4.5) they abandoned this form of analysis and postulated that a eutectic was adsorbed. This in itself seems unlikely and it would have

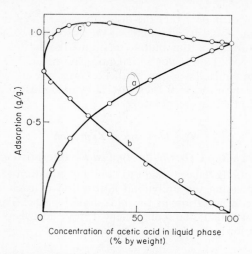

Fig. 4.3. Adsorption by charcoal from mixtures of acetic acid and water: (a) individual isotherm for acetic acid; (b) individual isotherm for water; (c) total adsorption.[13] (Reprinted from *Kolloid Zeitschrift* by permission of Dr. Dietrich Steinkopff Verlag, Darmstadt.)

been more profitable to question whether equation (4.5) expressed the correct relationship between adsorption and concentration.

The same basic assumption was independently taken further by Bartell,[16] who assumed that each component was adsorbed according to a Freundlich equation. In so far as the use of this equation could be justified, it seems more satisfactory to use it symmetrically with respect to the two components than for the "solute" only as had sometimes been done earlier. It led to the equation analogous to equation (4.5):

$$\frac{n_0 \Delta x}{m} = k_1 x^\alpha (1-x) - k_2 (1-x)^\beta x. \tag{4.6}$$

This equation was found to fit a number of composite isotherms, if a series of successive approximations were used to evaluate the four unknown constants k_1, k_2, α, and β. The isotherms related to adsorption from a variety of organic systems on charcoal and silica gel. An example of the fit obtained is shown in Fig. 4.4. Bartell did not consider the individual isotherms.

If this approach could be justified at all, it would be interesting to consider the use of a Langmuir equation instead of a Freundlich equation for the individual isotherms. The Langmuir equation at least has a more satisfactory theoretical basis in describing adsorption of gases. The equation for the composite isotherm would then be:

$$\frac{n_0 \Delta x}{m} = \frac{k_1 K_1 x (1-x)}{1 + k_1 x} - \frac{k_2 K_2 x (1-x)}{1 + k_2 (1-x)}. \tag{4.7}$$

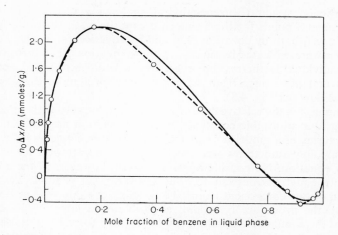

FIG. 4.4. Composite isotherm for adsorption by carbon from mixtures of benzene and ethyl alcohol.[16] Points, experimental; full curve according to equation (4.6). (Reprinted with permission from the *Journal of the American Chemical Society*.)

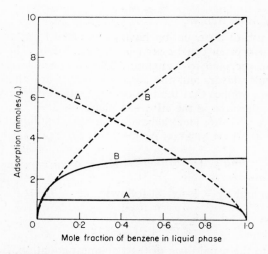

FIG. 4.5. Individual isotherms for adsorption by charcoal from mixtures of benzene and ethyl alcohol, calculated on the basis of the Freundlich equation (dotted lines) and the "Langmuir" equation (full lines): A, ethyl alcohol; B, benzene. (Reprinted with permission from *Nature*.)

For a particular system (benzene-ethyl alcohol-charcoal) it was found that the experimental points for the S-shaped composite isotherm could be fitted reasonably well both by equation (4.6) and by equation (4.7). The constants in each case were evaluated by successive approximations. The corresponding pairs of individual isotherms, however, are quite different from each other (Fig. 4.5) and also differ from those obtained by a more direct method described below (cf. Fig. 4.8).

This emphasizes the relative ease of fitting a simple curve with an equation containing four adjustable constants. Such an equation may have no significance in terms of a model for the physical system being considered. In the present case, pressure terms were replaced by concentration terms in the Langmuir equation. No consideration was given, however, to the relevance of applying equations derived for the adsorption of a single gas to the adsorption of one component from a liquid mixture containing at least two. In general, the procedure is invalid, as is discussed later in this chapter. The use of the Freundlich equation can be justified for very dilute solutions only (see Chapter 11). The use of an equation of the Langmuir type is described in the next section for an ideal case. This equation, however, has a slightly different form from that derived by Langmuir for the adsorption of single gases. Furthermore, a more complex equation is needed for any mixture which does not show ideal behaviour in the liquid phase.

A more fruitful approach is to consider the equilibria specifically involved in adsorption from the liquid phase.

(b) The Law of Mass Action

The correct use of an equation having the same form as the Langmuir equation for adsorption can be seen in a special case. The conditions are similar to those specified by Langmuir. The adsorbed layer is confined to a single molecular layer, and the adsorption sites are all equivalent. The second condition implies that the two components of a binary mixture have the same molecular area in the adsorbed phase.

For a mixture showing ideal behaviour in both the liquid phase and the adsorbed layer, the Law of Mass Action leads to[17]:

$$\frac{x_1^s x_2}{x_2^s x_1} = K. \tag{4.8}$$

As $x_2 = 1 - x_1$, and $x_2^s = 1 - x_1^s$, it follows that

$$x_1^s = \frac{K x_1}{1 + (K-1)x_1}. \tag{4.9}$$

This is similar to the usual form of the Langmuir equation. Because the two components have the same molecular area,

$$n_1^s + n_2^s = n^s, \tag{4.10}$$

where n^s is the total number of moles which can be accommodated in the adsorbed phase by unit weight of solid. Then:

$$x_1^s = n_1^s/n^s,$$

and

$$n_1^s = \frac{Kn^s x_1}{1+(K-1)x_1}.$$ (4.11)

The composite isotherm is therefore:

$$\frac{n_0 \Delta x}{m} = \frac{n^s x_1 x_2 (K-1)}{1+(K-1)x_1}.$$ (4.12)

The application of this equation to any set of experimental results can be tested by using it in the linear form:

$$\frac{x_1 x_2}{n_0 \Delta x/m} = \frac{1}{n^s}\left[x_1 + \frac{1}{K-1}\right],$$ (4.13)

from which n^s and K can be calculated. The further significance of K is considered in Chapter 14.

From the form of equation (4.12), and of (4.17) below, it can be shown that the U-shaped composite isotherm for systems showing ideal behaviour must be asymmetrical. The degree of asymmetry is indicated by the value of x_1 at which the isotherm reaches its maximum, and depends on K. This is discussed further in Chapter 14.

If the two components have different molecular areas, equation (4.8) has a different form (involving, as an extra parameter, the ratio of the molecular areas), and a much more complex expression than (4.11) is required to relate n_1^s to x_1. Moreover, mole fractions would have to be replaced by activities.

(c) Forms of the Langmuir Equation

Equation (4.8) can be re-written in the form

$$\frac{x_1^s}{1-x_1^s} = K\frac{x_1}{x_2}$$ (4.14)

and can be compared[17] with the Langmuir equation for adsorption of a single gas in the form:

$$\frac{\theta}{1-\theta} = Kp,$$ (4.15)

where θ is the fraction of the surface covered by the adsorbed gas.

The equations are analogous but are not identical in form. This is because the original Langmuir equation was derived so that θ should equal unity for $p = \infty$, whereas equation (4.8) is based on the requirement that x_1^s is unity when x_1 is unity. A device for comparing the equations is to postulate that the gas is adsorbed from a second component (vacuum) composed of "holes", which always has a mole fraction of unity in the gas phase.[17]

(d) A Modification of the Law of Mass Action

Without limiting the treatment to a specific model, Klinkenberg[18] assumed that adsorption is governed by an equation analogous to (4.8) but expressed in volume fractions. He then obtained

$$v_1^s = \frac{v^s \alpha V}{1+(\alpha-1)V},$$
(4.16)

where v_1^s is the volume of component 1 in the adsorbed layer on unit weight of solid, v^s is the total volume of this adsorbed layer, V is the volume fraction of component 1 in the liquid phase, and α is a constant. The corresponding equation for the composite isotherm (cf. 4.12) is

$$\frac{v_0 \Delta V}{m} = \frac{v^s V(1-V)(\alpha-1)}{1+(\alpha-1)V},$$
(4.17)

where ΔV is the decrease in the volume fraction of component 1 in the bulk liquid when adsorption occurs from a volume v_0 of original liquid on a mass m of solid.

Equation (4.17) can be tested by using the linear form:

$$\frac{V(1-V)}{v_0 \Delta V/m} = \frac{1}{v^s}\left[V + \frac{1}{\alpha-1}\right],$$
(4.18)

which is analogous to (4.13). Data for adsorption from mixtures of hydro-carbons by silica gel and alumina have been found to give good straight-line relationships when plotted according to equation (4.18). The use of the equation for adsorption from less ideal mixtures does not seem to have been investigated.

(e) The Nature of the Adsorbed Phase

For most real systems an equation more complex than (4.12) is needed to describe the composite isotherm, and the use of such an equation has not yet been developed. By the use of models of the adsorbed phase, however, it is possible to obtain individual isotherms for binary systems. Two models have commonly been used. In the first, the adsorbed phase is regarded as being confined to a single molecular layer. In the second (which is relevant only to adsorption by porous solids), the adsorbed phase is regarded as filling the available pore space.

(f) The Monomolecular Layer

A fundamental feature of this model is the recognition that the solid surface is completely covered by the adsorbed layer whatever the composition of the liquid phase.[19] The simplest further assumption which can be made is that the adsorbed layer is only one molecule thick. Thus one can refer to "monolayer adsorption" from solution, but the monolayer usually contains a mixture of the two components. Unless preferential adsorption is zero the molecular layer adjacent to the solid surface has a composition different from that of the bulk liquid, but all subsequent layers are part of the bulk liquid. The same assumption was investigated by Guggenheim and Adam[20] in applying the Gibbs equation to adsorption at the liquid–vapour interface (see Chapter 11).

It follows that:

$$n_1^s A_1 + n_2^s A_2 = A, \qquad (4.19)$$

where A_1 and A_2 are the partial molar areas occupied at the surface by the two components, respectively, and A is the surface area of unit weight of solid. A further assumption implicit in this equation is that the same area of the solid surface is potentially accessible to each component. This may readily be assumed for non-porous solids, but it is not necessarily valid when molecules of substantially different sizes are presented to a solid having narrow pores.

An alternative form of (4.19) was suggested independently, but on the basis of the same assumptions, by Kipling and Tester[7]:

$$\frac{n_1^s}{(n_1^s)_m} + \frac{n_2^s}{(n_2^s)_m} = 1, \qquad (4.20)$$

where $(n_1^s)_m$ and $(n_2^s)_m$ are the numbers of moles of the individual components, respectively, required to cover the surface of unit weight of solid completely. These values can be obtained by applying the Langmuir or B.E.T. equation to the isotherms for adsorption of the individual vapours by the solid, if (a further assumption) the orientations of the molecules are the same whether adsorbed from the vapour phase or from a liquid mixture. An equation equivalent to (4.20) was put forward by Williams[1] as applying approximately to the few results he obtained for the system charcoal–acetic acid–water. He did not, however, point out its significance.

It is important to note that terms of the Langmuir type as used in equation (4.7) are incompatible with this approach.[7] If they were used in equation (4.19), we should have:

$$\frac{A_1 k_1 K_1 x}{1 + k_1 x} + \frac{A_2 k_2 K_2 (1-x)}{1 + k_2 (1-x)} = A, \qquad (4.21)$$

which is a quadratic in x, and hence can, in general, give only two solutions for x. The experimental data, if interpreted in terms of monolayer coverage of a surface, require an equation which will give a number of solutions comparable with Avogadro's number.

Equation (4.20) could, in principle, apply to an adsorbed layer several molecules thick, in which all the layers had the same composition. This state of affairs is unlikely to exist in practice. If multilayer adsorption occurs, it seems more likely to involve a gradual change in composition from that of the first layer to that of the bulk liquid.

(g) The Pore-filling Model

The above approach could be called the "area-filling" concept, implying that the adsorbate completely fills the available surface of the adsorbent, whatever the concentration of the bulk liquid phase. Alternatively it has been held that adsorption by porous solids is essentially a "pore-filling" process, in which the available volume of the pores is the controlling factor.[21,22] Equation (4.19) is then replaced by:

$$n_1^s V_1 + n_2^s V_2 = V, \tag{4.22}$$

where V_1 and V_2 are the respective partial molar volumes of the adsorbates and V the pore volume of unit weight of adsorbent. The use of equation (4.22) similarly involves the assumption that the pores are equally accessible to the two components, i.e. that no molecular sieve action occurs.

The "pore-filling" theory may be of value in dealing with adsorption by solids which have relatively narrow pores. A major limitation, however, is that it provides no basis for comparing adsorption by porous and by non-porous solids, as it includes no mechanism by which the latter can occur.

Using the method described above (p. 40), Klinkenberg was able to show, for a porous solid, that the total volume of the adsorbed phase, v_t^s, was much less than the pore volume of the solid. For the systems studied, this is evidence against the "pore-filling" model. In one case, it was also possible to show, by combining the value of v_t^s with the specific surface area of the solid, that the thickness of the adsorbed layer was close to that of one molecule of the adsorbate being used. This supports the "area-filling" theory, provided that the assumptions underlying Klinkenberg's equation are accepted.

(h) Molecular Sieves

A special case of the pore-filling mechanism occurs with molecular sieves. Molecular sieve action (the sorption of one component to the complete exclusion of an other or others) has been extensively investigated for gaseous mixtures. It has generally been assumed that a similar process could occur in part of the pore structure of some adsorbents in contact with liquid mixtures.

A quantitative investigation has shown that complete molecular sieve action can occur in sorption from liquid mixtures by the Linde type of molecular sieve.[23] For such a situation equation (3.2):

$$\frac{n_0 \Delta x}{m} = n_1^s(1-x) - n_2^s x$$

becomes

$$\frac{n_0 \Delta x}{m} = n_1^s(1-x) \tag{4.23}$$

because n_2^s is always zero. Further, n_1^s is a constant for all values of x. Hence equation (4.23) describes a straight line which can be extrapolated to $x = 0$, to give n_1^s, the pore volume of the solid for the component in question. Examples of this are shown in Fig. 4.6. The extrapolated values are 0·164 c.c./g. for n-butyl chloride and 0·170 c.c./g. for water.

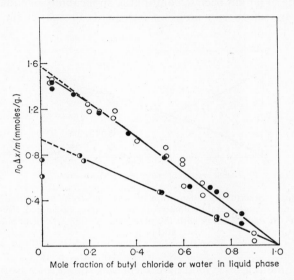

FIG. 4.6. Adsorption from liquid mixtures by Linde Molecular Sieve 5A: ●, n-butyl chloride and benzene at 20°c; ○, n-butyl chloride and carbon tetrachloride at 20°c; ◑, water and furfuryl alcohol at 30°c. The values on the ordinate are for butyl chloride; they should be multiplied by 10 for water.[23] (Reproduced with permission from the *Transactions of the Faraday Society*.)

Complete molecular sieve action has been found in adsorption by Linde molecular sieve 5A from mixtures of benzene and n-hexane; hexane is sorbed and benzene is excluded.[24] The difference in the effective molecular diameters (4·9 and 6·3 Å, respectively) is small enough to emphasize that very careful attention must be given to the sizes both of molecules and of pore openings when adsorption by porous solids is being considered.

It is probably reasonable to predict whether molecular sieve action will occur from solutions with this type of adsorbent either from the relative diameters of the molecules concerned and the pore openings of the solid, or from a knowledge of the extent to which the respective vapours are sorbed by the solid. For molecular sieves, the treatment described in the preceding section can normally be expected to be adequate.

For other porous solids, Williams' method can undoubtedly be used to determine the total quantity of each component held in the pores. It is not yet clear, however, whether it can be refined sufficiently for a distinction to be made between primary adsorption and mechanical filling of the remaining volume in the pores by the bulk liquid ("capillary" adsorption); this is particularly important for adsorbents in which a considerable percentage of the pore volume is in wide pores.

Its main value so far has been to confirm the applicability of equation (4.20) to adsorption by solids with pores so narrow as to restrict the adsorbate

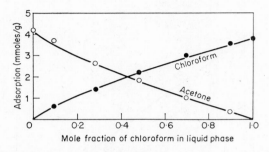

FIG. 4.7. Individual isotherms for adsorption by charcoal from mixtures of chloroform and acetone.[8] Points: experimental (Williams' method); curve: calculated. (Reprinted from the *Journal of the Chemical Society* with the permission of The Chemical Society and the author.)

essentially to a monolayer. Good agreement was found for adsorption from a number of organic mixtures by an activated coconut shell charcoal.[8] A typical case is shown in Fig. 4.7. When this method is applied to the system benzene–ethyl alcohol–charcoal, it gives individual isotherms (Fig. 4.8) which differ from those of Fig. 4.5, obtained by using equations of the

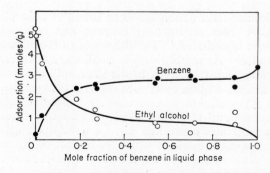

FIG. 4.8. Individual isotherms for adsorption on charcoal from mixtures of benzene and ethyl alcohol; curves calculated by use of equation (4.20); points obtained experimentally by Williams' method. (Reprinted with permission from the *Journal of the Oil and Colour Chemists' Association*.)

Freundlich or Langmuir type for the adsorption of the individual components.[7] It may be significant that, for low concentrations of each component, the isotherms of Fig. 4.8 agree much better with those derived from a Langmuir expression than those derived from a Freundlich expression.

(i) Present Use of the Above Methods

Adsorbents can be classified into four groups: non-porous solids, solids with pores which are wide relative to molecular diameters, solids with narrow pores, and molecular sieves. The grouping is not rigid, but is frequently convenient. In some cases the surfaces of solids in the second category may seem difficult to distinguish from the free surfaces of non-porous solids. Similarly, if a solid in the third category has a spread of pore sizes, the narrowest pores might be expected to exhibit molecular sieve action in the presence of some liquid mixtures.

In the absence of further development of the experimental approach, three attitudes to the determination of individual isotherms have become apparent.

(i) Many composite isotherms have been published without a corresponding analysis to give the individual isotherms. This may imply that the authors concerned consider that there is no reliable method of analysis.

For non-porous solids, no systematic examination of Williams' method seems to have been attempted.

(ii) For adsorption by porous solids, equation (4.22) has been used,[22] or its use has been advocated[21] because it is believed that adsorption from solutions of the type discussed in this chapter is multimolecular, unless this is prevented by the narrowness of the pores. On this view, no analysis can be made for adsorption on non-porous surfaces until an equation has been developed to relate n_1^s and n_2^s when multilayer adsorption occurs.

(iii) The possibility of using equation (4.20) for adsorption both by porous and non-porous solids has been explored, in the belief that, for many systems, it is a good approximation to assume that adsorption is confined to a monolayer. For some cases this analysis fails and it is then necessary to assume that adsorption extends to two or more molecular layers. A definite proof of monolayer adsorption is not, in general, obtainable, but the circumstantial evidence accumulated by the attempted use of this method should ultimately lead to an assessment of its value. At present it appears to be the only general method for attempting to calculate individual adsorption isotherms at any type of interface.

In particular cases, individual isotherms are obtained by alternative methods in different laboratories. The choice of method turns largely on whether or not monolayer adsorption from solution is considered to be possible.

MONOLAYER† AND MULTILAYER ADSORPTION

General Considerations

The need to discuss this topic arises because, for adsorption from many completely miscible pairs of liquids, it has been possible to analyse the composite isotherm into individual isotherms by using the monolayer hypothesis. This does not prove that adsorption is confined to a monolayer, but leaves the possibility open. On the other hand, in some cases multilayer adsorption undoubtedly occurs.

The certain occurrence of multilayer adsorption can be recognized as follows. The highest value which $n_0 \Delta x/m$ can have, compatible with monolayer adsorption, corresponds to complete coverage of the surface by component 1, and no adsorption of component 2. Equation (3.2) then reduces to:

$$n_0 \Delta x/m = (n_1^s)_m (1-x). \qquad (4.24)$$

If, therefore, for any value of x, the value of $n_0 \Delta x/m$ exceeds that of $(n_1^s)_m (1-x)$, the composite isotherm cannot be resolved on the monolayer hypothesis. Systems in which this occurs include the adsorption by carbon black from mixtures of ethanol and n-dodecane[25] at 25°C, and on Graphon (a graphitized carbon black) from mixtures of methyl alcohol and benzene,[26] at 20°C. The significance of these temperatures is discussed later (Chapter 10).

In cases to which the above situation does not apply, the monolayer theory was first used extensively by Kipling in investigating adsorption by charcoal.[7,8] This was criticized by Jones and Mill[22] on the ground that adsorption from the liquid phase corresponded to adsorption from a mixed saturated vapour, and that in general this would involve capillary condensation if the solid were porous. Tryhorn and Wyatt's evidence (p. 11), however, is that the capillary condensate from a saturated mixed vapour has the same composition as that of the liquid with which it is in equilibrium. What matters, therefore, is the number of layers present in the adsorbate prior to the onset of capillary condensation.

Hansen and Hansen's criticism[21] was essentially similar, but was reinforced by a further argument concerning the adsorbent. The isotherms for the adsorption of single vapours on the charcoal used in this work were of Type I in Brunauer's classification,[27] except for a small sharp rise in adsorption at pressures close to saturation. A Type I isotherm was usually

---

† The terms "monolayer adsorption" and "multilayer adsorption" are used here as they are applied in the restricted sense to adsorption of vapours. The former means that the adsorbate is confined to one molecular layer, the latter means that it extends to more than one molecular layer. In the less restricted sense, a possible confusion should be avoided. As most experiments on adsorption of vapours are carried out with single components and not with mixtures, the term "a monolayer (of adsorbate)" refers to a homogeneous monolayer. In adsorption from solution, the monolayer usually contains both components. In this book, therefore, the monolayer of material adsorbed from solution is normally a "mixed" monolayer except where the context shows that a different sense is intended.

regarded as implying monolayer adsorption, but it had been suggested[28] that even at low pressures, this type of isotherm was compatible with the occurrence of capillary condensation simultaneously with monolayer adsorption. This implies the same point as was raised by Jones and Mill, and can be met in the same way. It goes further, however, because the mono-layer values, $(n_1^s)_m$ and $(n_2^s)_m$, used in equation (4.20), were obtained by assuming that the vapours were adsorbed singly by a monolayer mechanism to which the Langmuir treatment was applicable. The numbers so calculated would not, of course, have this value if monolayer adsorption were accom-panied by capillary condensation at low pressures. This argument appears to apply mainly, though not exclusively, to narrow-pored charcoals, about which discussion continues. It is still claimed that, in charcoals of this type, most of the pores are too narrow to admit of capillary condensation,[29] unless the measured "pore diameters" are really the diameters of the narrow openings to much wider "caves".

Hansen and Hansen also calculated[21] that for adsorption from the system benzene–ethyl alcohol, the total volume of adsorbate, expressed as a volume of liquid, was almost constant for the whole range of composition of the liquid mixtures. This supported the pore-filling mechanism. On the other hand, in adsorption from the system water–pyridine on the same charcoal, the adsorption of water fell from 7 to 0 mmoles/g. over a range in concen-tration at which adsorption of pyridine remained almost constant at 4 mmoles/g. This is not in accordance with the pore-filling mechanism, but can be explained by a modification of the monolayer mechanism[8] (see p. 55).

Schay and Nagy's Analysis

Evidence supporting the hypothesis that adsorption may be confined to a single molecular layer is to a large extent circumstantial. Further support is provided by an analysis proposed by Schay and Nagy.[30-32] They point out that some composite isotherms have a substantially linear section (Fig. 4.9). Over such a section, the equation

$$\frac{n_0 \Delta x}{m} = n_1^s(1-x) - n_2^s x$$

$$= n_1^s - (n_1^s + n_2^s)x \qquad (4.25)$$

defines a straight line. Schay and Nagy consider that this is most probably due to n_1^s and n_2^s remaining constant over that section of the isotherm, i.e. that for a considerable range of x, the composition of the adsorbed phase remains constant. This composition can be calculated if the linear section of the isotherm is extrapolated, giving values of $(n_1^s)_c$ when $x = 0$ and $-(n_2^s)_c$ when $x = 1$. (A similar analysis can be made if the isotherm is expressed in other units, e.g. $w_0 \Delta c/m$.) If appropriate molecular areas are assigned to the two components, a specific surface area can be calculated for the solid. Comparison of this value with a value determined by a standard

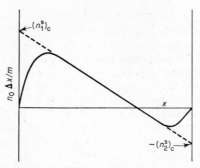

FIG. 4.9. Composite isotherm showing substantial linear section.

method (e.g. low-temperature adsorption of nitrogen) shows whether it is reasonable to assume that the adsorbed phase is confined to a single molecular layer. More generally, the mean molecular thickness of the adsorbed phase, t (as a number of molecular diameters), can be calculated from the equation:

$$t = \frac{(n_1^s)_c}{(n_1^s)_m} + \frac{(n_2^s)_c}{(n_2^s)_m} \tag{4.26}$$

if the monolayer values $(n_1^s)_m$ and $(n_2^s)_m$ are known for the two components (e.g. from adsorption data for the single vapours).

The conditions envisaged by Schay and Nagy seem at first unlikely to be fulfilled exactly in very many cases, as can be seen from thermodynamic considerations. Thus

$$\mu_1^s = (\mu_1^s)^0 + RT \ln (f_1^s x_1^s) \tag{4.27}$$

and

$$\mu_1^l = (\mu_1^l)^0 + RT \ln (f_1^l x_1^l), \tag{4.28}$$

where μ_1 and $(\mu_1)^0$ are the chemical potentials of component 1 in a given mixture and in the standard state, respectively, f_1 being the activity coefficient; the superscript l refers to the bulk liquid. At equilibrium

$$\mu_1^s = \mu_1^l + A_1 \gamma, \tag{4.29}$$

where γ is the interfacial tension, and A_1 is the partial molar area of component 1. From the relationship between γ and $f_1^l x_1^l$ given by the Gibbs equation, it follows that when x_1^l changes, x_1^s can only remain constant if f_1^s changes precisely so as to counter the change in $f_1^l x_1^l$. This cannot be expected to occur frequently.

The restriction does not, however, apply if a saturated adsorbed phase is formed over an appreciable range of concentration in the bulk liquid. An extreme example is sorption by a molecular sieve from a mixture of two liquid components, only one of which can enter the cavities (p. 42); this produces an almost linear isotherm. The example is relevant, however, only to examining the behaviour of saturated adsorbed phases and not to the determination of the molecular thickness. A more relevant case would be the formation of a complete monolayer by one of the two components, which might be expected to occur in the adsorption of solids from some solutions.

Another possibility arises if the solid surface is heterogeneous (and few, if any, are not).[33] This is illustrated in Fig. 4.10 for a surface formed from two kinds of site. One is readily saturated with one component, and the other

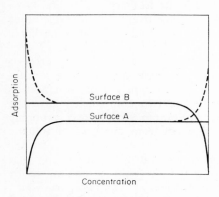

FIG. 4.10. Individual isotherms for adsorption on a heterogeneous surface.[33] (Reproduced with permission from the *Transactions of the Faraday Society*.)

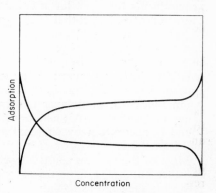

FIG. 4.11. Individual isotherms with a central section approximately constant.[33] (Reproduced with permission from the *Transactions of the Faraday Society*.)

with the second component. In the middle region of concentration, the adsorbed phase is of constant composition and can be analysed by Schay and Nagy's method.

Alternative interpretations of the linear section of the composite isotherm can be sought by examining the conditions necessary for $\dfrac{d^2(n_0 \Delta x/m)}{dx^2}$ to be zero. These are generally complex. An approximation to these conditions can, however, occur, if

$$n_1^s = a_1 + b_1 x \tag{4.29}$$

and

$$n_2^s = a_2 + b_2(1-x) \tag{4.30}$$

over the given range of concentration, b_1 and b_2 being small, so that the individual isotherms have the form shown in Fig. 4.11. The composite isotherm cannot in such cases be strictly linear, but the deviation from linearity should be small and application of the Schay and Nagy analysis should give an approximate value for t. This is unlikely to be greatly in error because the composite isotherm should be noticeably non-linear if either b_1 or b_2 is large.

Liquid mixtures which are far from ideal are likely to give rise to this situation. Thus for the system methanol–carbon tetrachloride at 25° both the vapour phase and the adsorbed phase on charcoal are of almost constant composition over a considerable range of liquid composition.[9] Similar behaviour is shown by the system nitromethane–carbon tetrachloride with silica gel as adsorbent.[22] Such systems do not fulfil the thermodynamic requirements for strict linearity of the composite isotherm. They are, however, so far from ideal that $f_1^l x_1^l$ changes very little over a considerable range of x_1^l; consequently x_1^s can remain almost constant without any remarkable compensating changes in f_1^s.

A special case arises if $b_1 = b_2$; whether the constants are large or small, the isotherm must then be strictly linear over the range of concentration for which equations (4.29) and (4.30) apply. This situation is compatible with monolayer adsorption if the two components have the same molecular area when adsorbed. An example of this is probably found in adsorption by charcoal and silica gel from mixtures of toluene and iso-octane.[34]

Most of the systems quoted by Schay, either from his own work or that of other authors, can be treated either by the approximate method or by regarding the solid as having a heterogeneous surface, with two parts which can be saturated respectively by the two liquid components. On this basis his results show that for adsorption from a number of systems on charcoal and alumina, the adsorbate is confined to a single molecular layer, within a possible error of 5%.

Similar results have been obtained for adsorption by non-porous carbon blacks,[33] which shows that restriction of adsorption to a single molecular layer does not result simply from the restricted space available in some porous adsorbents. Figure 4.12 shows what is probably an example of adsorption on a markedly heterogeneous surface, that of the carbon black, Spheron 6. It seems likely that the linear range corresponds to saturation of the polar sites by cyclopentanone and of non-polar sites (existing in relatively small proportion) by n-heptane. The very different isotherm obtained for adsorption on Graphon (on which the number of polar sites is negligible) tends to confirm this interpretation (Fig. 4.13).

The conclusions which have been drawn from this analysis apply only to systems which give isotherms with a linear section. These are in a minority. They are, however, generally systems which are far from ideal in the liquid state. If adsorption is, nevertheless, confined to a single molecular layer, even for these systems, it seems unlikely that systems nearer to ideal behaviour will be found to exhibit multimolecular adsorption.

For regular solutions a treatment based on statistical mechanics shows that adsorption is confined essentially to a single layer, provided that the system is well above the temperature at which phase separation occurs;[35] T_c is defined as the temperature at which an equimolar mixture just separates into two phases. The number of molecular layers needed to account for

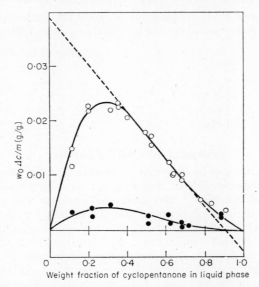

FIG. 4.12. Composite isotherm for adsorption from mixtures of cyclopentanone and n-hexane onto carbon blacks; ○, on Spheron 6; ●, on Graphon.[33] (Reproduced with permission from the *Transactions of the Faraday Society*.)

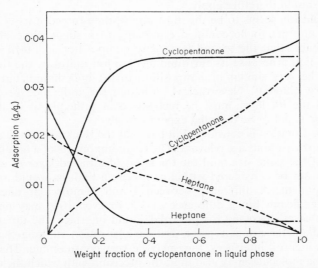

FIG. 4.13. Individual isotherms for adsorption from mixtures of cyclopentanone and n-heptane; full line, adsorption on Spheron 6 (dotted lines show suggested saturation of polar sites by cyclopentanone and of non-polar sites by n-heptane); broken lines, adsorption on Graphon.

99% of the difference in composition between the surface layer and the bulk phase is then dependent on the temperature, T, as follows:

T/T_c	number of layers
10	1·3
5	1·6
1·5	3·5
1·01	23
1·00	∞

Adsorption from Mixed Vapours

More direct evidence can be obtained by examining adsorption from mixed vapours, as it is then possible to determine the composition of the adsorbed phase. Cines and Ruehlen[36] measured the isotherm for adsorption by silica gel from mixtures of benzene and 2,4-dimethyl pentane; the temperature of 65·6°C was chosen because the two components have the same vapour pressure at that temperature. By desorbing the adsorbate after equilibrium was established, they were able to determine its composition. Multilayer adsorption of the mixed vapour was observed, with some preferential adsorption of benzene. This preferential adsorption persisted beyond the first layer, but it was concluded that "selectivity is vested primarily in the monolayer". For such a system, the observed situation should be very close to that obtaining at the liquid–solid interface, to which the same conclusion should therefore apply.

This conclusion seems to be the most appropriate for most cases in which adsorption appears to be confined essentially to a single molecular layer. It is improbable that the second layer is of composition identical with that of the bulk liquid because its surroundings are not uniform; the third layer is likely to be very similar in composition to the bulk liquid, but the first layer is very different. Nevertheless, although the difference between the second and the ith layer must be recognized in principle, it is at present generally too small to be detected experimentally.

When this situation is considered in terms of the thickness of the adsorbed layer, two further points are relevant. The two components of a liquid mixture are rarely of the same size, and the thickness of a mixed layer of adsorbate should thus not be considered within as precise a limit as would apply to a single component. A difference between 1 and 2 molecular diameters is normally significant, but a difference of 1% or even 10% may not be. In most of the cases discussed for the solid–liquid interface, the differences in molecular diameter are not great, but in adsorption of long-chain compounds at water–oil interfaces, they may be very considerable. The thermal motion of the molecules may also limit the precision with which the thickness of the adsorbed layer can be defined. Again, this may be more important at the liquid–liquid and liquid–vapour interfaces than at the liquid–solid interface where the surface is rigid.[37]

The Potential Theory

Polanyi modified the potential theory which had been introduced for adsorption of gases[38] to deal with adsorption from solutions of sparingly soluble solutes.[39] The significance of this approach is that it is not necessary to postulate a specific thickness for the adsorbed layer, adsorption being considered as a function of the adsorption potential at any given distance from the adsorbing surface. For a single gas, this potential, ε, is equal to the work done in bringing one mole of adsorbate from infinity to a specified distance from the surface. For a sparingly soluble solute, Polanyi gave the equation:

$$\varepsilon_1 = RT \ln \frac{c_0}{c} + \varepsilon_2 \frac{V_1}{V_2} \qquad (4.31)$$

or

$$V_2 \varepsilon_1 - V_1 \varepsilon_2 = V_2 RT \ln \frac{c_0}{c}, \qquad (4.32)$$

where ε_1 is the adsorption potential of the solute and ε_2 that of the solvent for a solution of concentration c_1; c_0 is the limiting solubility for the system, and V_1 and V_2 are the molar volumes of solute and solvent respectively.

Little use has been made of this theory, but an attempt has been made to extend it for use with completely miscible pairs of liquids.[40] This gives, for the composite isotherm:

$$\frac{V \Delta c}{m} = \int_0^\infty \left(\frac{x_{1\phi}}{V_\phi} - \frac{x_{1B}}{V_B} \right) d\phi, \qquad (4.33)$$

where $x_{1\phi}$ is the mole fraction of component 1 in the adsorbed phase at an equipotential surface approximately parallel to the surface of the adsorbent and enclosing a total volume of adsorbate, ϕ; x_{1B} is the mole fraction of component 1 in the bulk phase, and $\bar{V}_\phi$ and $\bar{V}_B$ are the molar volumes for the mixture in the adsorbed phase and in the bulk solution, respectively. With three simplifying assumptions, the equation

$$\frac{a_{1\phi}}{a_{2\phi}^\alpha} = \frac{a_{1B}}{a_{2B}^\alpha} e^{[\varepsilon_1(\phi) - \alpha\varepsilon_2(\phi)]/RT} \qquad (4.34)$$

can be obtained for the corresponding activities, α being the ratio of the partial molar volumes of components 1 and 2 in the adsorbed phase. Equation (4.34) is used to solve equation (4.33) by assuming that $\dfrac{a_{1\phi}}{a_{2\phi}^\alpha}$ is the same function of $x_{1\phi}$ as $\dfrac{a_{1B}}{a_{2B}^\alpha}$ is of x_{1B}. The values of $\varepsilon_1(\phi)$ and $\varepsilon_2(\phi)$ are obtained from the isotherms for adsorption of the corresponding vapours.

Equation (4.33) has been tested for adsorption from n-propanol–water mixtures by Spheron 6 (1000°c).† The results are shown in Fig. 4.14. The curve given by the simple theory fits the experimental points less satisfactorily than does a modified curve based on the assumption that very strong interaction between water and the surface of the adsorbent occurs, so that water is adsorbed as a layer of oriented dipoles.

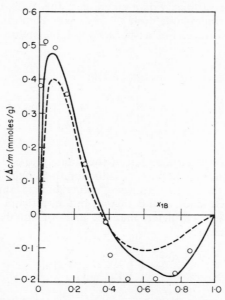

FIG. 4.14. Adsorption from n-propanol–water mixtures by Spheron 6 (1000°c) at 25°c; points, experimental; broken line, according to simple potential theory; full line, according to modified potential theory.[40] (Reprinted from the *Journal of Physical Chemistry* with the permission of the American Chemical Society.)

Aqueous systems may well be thought to provide the most stringent test of any theory of this kind. It remains to be seen whether this approach proves successful in providing an adequate theory of adsorption from solution, especially as the potential theory has recently been revived for adsorption of gases, with the possibility of calculating ε as a function of ϕ theoretically.

Special Cases

A few cases are known in which the composite isotherm has a linear section, and the adsorbed layer is two molecules thick, e.g. adsorption by silica gel from mixtures of alcohol and benzene.[41] In this case the first layer is a

† This adsorbent is made by heating the carbon black, Spheron 6, to 1000°c in an inert atmosphere.

chemisorbed layer of alcohol, and the second layer is a physically adsorbed mixture.

In adsorption from some aqueous systems, e.g. pyridine–water,[8] the composite isotherm (Fig. 4.15) is convex to the concentration axis over part of the range, suggesting that the adsorbed phase has an excess of water or a deficiency of pyridine relative to what would be expected with a normal U-shaped isotherm. This is confirmed in attempts to calculate individual isotherms according to the monolayer theory, as curves showing a maximum

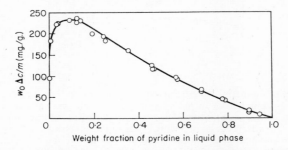

FIG. 4.15. Composite isotherm for adsorption on charcoal from mixtures of pyridine and water.[8] (Reprinted from the *Journal of the Chemical Society* with the permission of The Chemical Society and the author.)

and a minimum are obtained. It would be possible to attribute this effect to generalized multilayer formation or to pore-filling. The latter is ruled out when the individual isotherms for adsorption from the mixed vapours are obtained. These show (Fig. 4.16) a sharp fall in the amount of water adsorbed as concentration increases, with virtually no corresponding increase in adsorption of pyridine. The effect, moreover, cannot be attributed to molecular sieve action.

A more specific proposal is that the adsorbed pyridine is hydrated to extents which vary with the composition of the bulk liquid. If the degree of hydration is assumed to follow a simple Law of Mass Action expression, a constant can be chosen empirically and calculations of individual isotherms can be made from the data for the composite isotherm. These fit the data obtained for adsorption from the mixed vapours (Fig. 4.16). In this treatment it is assumed that water is adsorbed, when free from pyridine, as a dimeric molecule. This situation is intermediate between simple monolayer adsorption and generalized multilayer adsorption. For purposes of calculation it is convenient to regard it as a special case of monolayer adsorption, in which complex molecules are adsorbed, especially because there is no evidence for the existence of pyridine in a second layer.

An unusual isotherm is shown in Fig. 4.17. Although steps have been observed in isotherms for adsorption from solutions of solids (Chapter 7), they are rarely found in adsorption from completely miscible pairs of liquids.

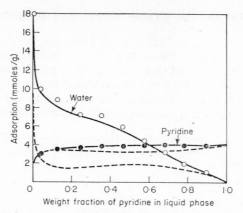

FIG. 4.16. Individual isotherms for adsorption on charcoal from mixtures of pyridine and water. Points: adsorption from mixed vapours. Broken curves: calculated from liquid-phase adsorption on the assumption that water is adsorbed as monomer and without interaction between adsorbed components. Full curves: calculated from liquid-phase adsorption on the assumption that water is adsorbed as dimer, and as hydrated pyridine.[8] (Reproduced with permission from the *Journal of the Chemical Society.*)

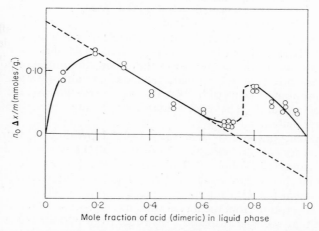

FIG. 4.17. Adsorption on Spheron 6 (1000°C) from mixtures of n-butyric acid and cyclohexane[33] at 20°C. (Reproduced with permission from the *Transactions of the Faraday Society.*)

The first part of the isotherm corresponds, according to the Schay analysis, to a mixed monolayer, and the step is probably due to the formation of a second layer containing a high proportion of butyric acid.[33]

The conditions which favour respectively monolayer and multilayer formation are discussed in more detail in Chapter 5.

TYPES OF COMPOSITE ISOTHERM

Simple Classification

It is useful at this stage to classify the types of composite isotherm which are met in practice as this will help the discussion of individual isotherms. A considerable range of adsorbates and adsorbents has now been examined, and a simpler classification can be put forward than that of Ostwald and de Izaguirre. Basically there are three types of composite isotherm, if they

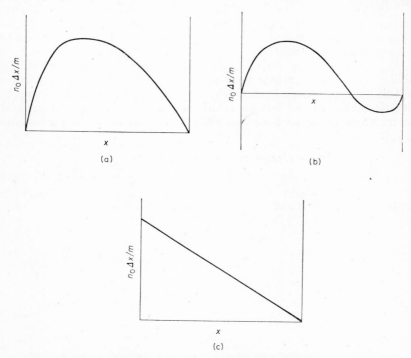

FIG. 4.18. Types of composite isotherm: (a) U-shaped; (b) S-shaped; (c) linear.

are classified according to shape. The characteristic features are shown in Fig. 4.18:

(a) a one-branch or U-shaped isotherm,
(b) a two-branch or S-shaped isotherm, the two branches usually being of different size,
(c) a linear isotherm.

The first two are very common and will be discussed in detail later (Chapter 10). Type (a) may now replace the first in Ostwald and de Izaguirre's classification, and type (b) may replace almost all of the remainder; in no case, however, is their breakdown into individual isotherms necessarily appropriate.

Type (c), corresponding to their case of adsorption of solvent only, has so far been found in special circumstances, but may become of increasing importance. It occurs in sorption by molecular sieves when one component can enter the pores and the other cannot (p. 43). In such cases the composite isotherm is probably not strictly linear, as normal competitive adsorption can occur on the external surfaces in addition to sorption by the pores. Thus the composite isotherm is the combination of two, one being of type (c) and the other of type (a) or (b). In practice, the adsorptive capacity of the external surface is so small compared with the capacity of the pores, that the deviation from linearity is very slight.

Sorption of this kind by molecular sieves may hardly be regarded as involving an interface as that term is usually understood. It is introduced here because a partial molecular sieve activity may be expected to operate with some solids having fine pores distributed over a range of diameters. It is therefore useful to have examples of this behaviour in the extreme cases.

A More Detailed Classification

A finer classification can be made by recognizing variants of the basic shapes (a) and (b). Nagy and Schay[30,31] recognize five types (Fig. 4.19); three correspond to (a) and two to (b) as illustrated above.

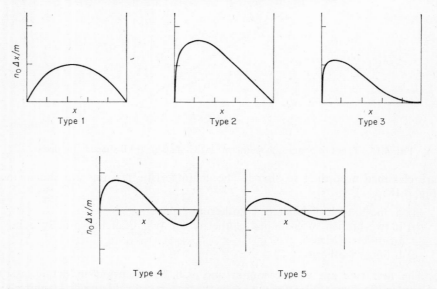

FIG. 4.19. Types of composite isotherm according to Schay.[31] (Reproduced with permission from *Periodica Polytechnica*.)

TYPES OF INDIVIDUAL ISOTHERM

When the experimental data are susceptible of analysis according to the monolayer hypothesis, two main types of individual isotherm are obtained. In the first (Fig. 4.20(a)), there is no point of inflexion in the curves. In the second (Fig. 4.20(b)), a point of inflexion is found. It follows, from the area-filling theory used, that if one curve of the pair has a point of inflexion, the other must have a corresponding inflexion; if one is without inflexion, the other must also be without inflexion. The first type of individual isotherm

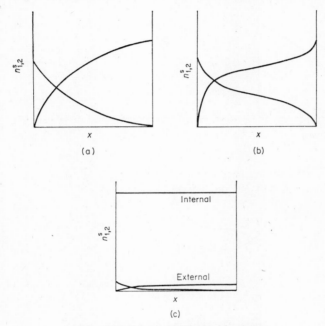

FIG. 4.20. Types of individual isotherm.

corresponds to the U-shaped composite isotherm, the second to the S-shaped composite isotherm. Further reference to these types of isotherm will be found in Chapter 10.

The first type of individual isotherm may vary from being almost linear, when there is no strong preferential adsorption (e.g. adsorption from mixtures of ethylene dichloride and benzene by silica gel[41]) to a strongly curved isotherm almost conforming to the Langmuir equation, when preferential adsorption is marked (e.g. adsorption from mixtures of benzene and cyclohexane by charcoal[8]). The three variants in Nagy and Schay's classification are shown in Fig. 4.21. (In the third type the individual isotherms have points of inflexion because the composite isotherm, although it has no negative branch, does itself have a point of inflexion.)

The second type of individual isotherm is always inflected and usually strongly curved (e.g. adsorption from mixtures of ethyl alcohol and benzene by charcoal[7]) though some depart only slightly from linearity (e.g. ethylene dichloride and benzene with charcoal[7]). These are typified by the fourth and fifth pairs of isotherms in Fig. 4.21.

The third type of individual isotherm is obtained when total molecular sieve action occurs. In this case the concept of monolayer or multilayer adsorption in the pores is irrelevant as they can accept only one component.

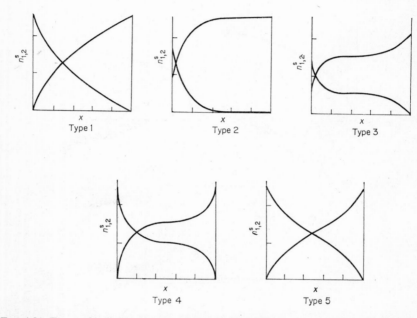

FIG. 4.21. Types of individual isotherm according to Schay.[31] (Reproduced with permission from *Periodica Polytechnica*.)

Strictly, allowance should be made for the competitive adsorption which must occur on the external surface of the sorbent, but the absolute magnitude of this is so small that it can generally be ignored. An attempt to represent this approximately is shown in Fig. 4.20(c).

THE EFFECT OF TEMPERATURE

From a small number of investigations it can be seen that temperature affects several factors which are important in adsorption from solution. For completely miscible systems, selective adsorption generally decreases with rise in temperature.[42] This now seems to be generally observed, whether the composite isotherm is U-shaped or S-shaped. It is the expected effect for an exothermic process, and is compatible with a decrease in magnitude, and hence in specificity, of van der Waals forces as the temperature increases. Thus the preferential adsorption by charcoal of acetic acid from aqueous solutions decreases as the temperature rises.[43] For polar solids the effect

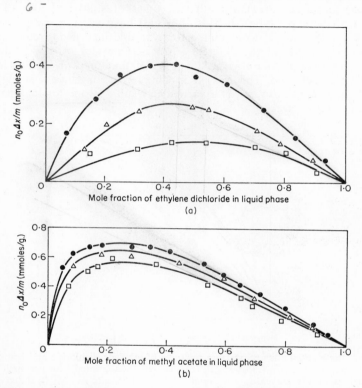

Fig. 4.22. Effect of temperature on adsorption by alumina (boehmite) from mixtures of (a) ethylene dichloride and benzene, (b) methyl acetate and benzene[44]: ●, 20°c; △, 40°c; □, 60°c.

appears to be more marked for simple mixtures (e.g. ethylene dichloride–benzene–alumina, Fig. 4.22(a)) than for those in which one component is more polar than the other (methyl acetate–benzene–alumina, Fig. 4.22(b)).[44]

Rao[42] observed that there was no significant change with temperature in the magnitude of selective adsorption by silica gel from benzene–ethyl alcohol mixtures. This has also been confirmed and has been attributed to chemisorption of the alcohol by the silica gel.[41]

FORMS OF ISOTHERM SUITABLE FOR COMPARING ADSORPTION AT DIFFERENT INTERFACES

If the above individual isotherms (or composite isotherms) are used for comparing adsorption by different solids, the comparison is made on the basis of two factors: the inherent capacity of the solid surface for selecting one component in preference to the other, and the total extent of the surface (the specific surface area). As different types of solid and also different samples of the same type (such as charcoal) have varying specific surface areas, it is often useful to eliminate this variable in order to compare the inherent selectivities of the different surfaces. One way of doing this is to express the adsorption isotherms in terms of unit area of surface instead of unit weight of solid. This is done in Chapter 11, in which it is suggested that the unit, μmoles/sq. m., is a particularly suitable one for making comparisons of all interfaces.

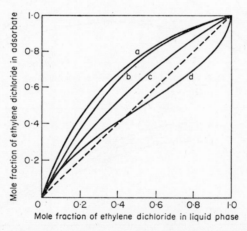

FIG. 4.23. Preferential adsorption from mixtures of ethylene dichloride and benzene by: (a) boehmite, (b) γ-Al$_2$O$_3$, (c) silica gel, (d) charcoal.[45] (Reprinted with permission from "Proceedings of the Second International Congress of Surface Acitivity", Butterworths.)

An alternative method is to represent competitive adsorption by plotting the mole fraction of one component in the adsorbed phase against its mole fraction in the equilibrium phase.[6] The diagonal then represents zero preferential adsorption. The extent of departure from the diagonal represents the extent of inherent selectivity shown by any solid.[45] Examples are shown in Fig. 4.23.

CHEMISORPTION

Adsorption from mixtures of simple organic compounds such as have been described so far is usually expected to be physical adsorption at room temperature. It has been shown, however, that even in adsorption at room temperature, chemisorption is sometimes possible.

In adsorption on alumina (boehmite, γ-AlO.OH) from mixtures of ethyl alcohol and benzene, the maximum in the composite isotherm corresponded to more than monolayer adsorption,[44] as shown by the methods described earlier in this chapter. This might have been due to multilayer adsorption or to a pore-filling process, but other observations suggested that chemisorption had to be considered. First, equilibrium was established much more slowly than with other systems (Fig. 4.24); whereas equilibrium with such adsorbents is normally established well within 24 hours with simple adsorbates, a drift was observed in this case over several days, and it was

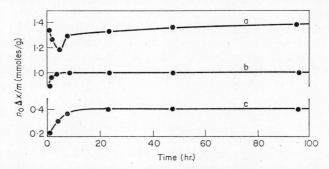

FIG. 4.24. Rates of adsorption from liquid mixtures by alumina, as shown by variation of change in concentration with time: (a) ethyl alcohol–benzene with boehmite, at 0·318 mole fraction of ethyl alcohol; (b) ethyl alcohol–benzene with a surface of the gibbsite type, at 0·322 mole fraction of ethyl alcohol; (c) ethylene dichloride–benzene with boehmite, at 0·331 mole fraction of ethylene dichloride.[44] (Reproduced with permission from the *Journal of the Chemical Society*.)

necessary to allow a period of three weeks to elapse to ensure that equilibrium conditions had been established. This is consonant with slow chemisorption, altering the surface of the solid, together with physical adsorption on the altered surface. The existence of a slow irreversible adsorption of ethyl alcohol by boehmite was established in separate experiments with the vapour.[46,47] The chemical reaction can be represented as:

$$\text{AlO.OH} + \text{EtOH} \rightarrow \text{Al(OH)}_2\text{(OEt)}.$$

The course of the adsorption process from the liquid phase can be represented as physical adsorption initially on the fresh surface, followed by slow chemisorption of the alcohol, and the re-establishment of competitive physical adsorption on the "alkylated" surface. The dip in the curve of $n_0 \Delta x/m$ against time (Fig. 4.24) probably represents displacement of some of the alcohol, initially adsorbed physically, as chemisorption proceeds; the ethyl alcohol molecule occupies a greater area when chemisorbed than when physically adsorbed.[46]

For periods of adsorption of one day, raising the temperature of the system from 20° to 60°C has less effect than is expected for physical adsorption alone (see above). Moreover, if a system in which physical adsorption only occurs (e.g. ethylene dichloride–benzene–boehmite) is held for successive periods of one day at 20°, 60°, and 20°C, the liquid phase has the same composition at the end of the third day as at the end of the first. If this is done with the system ethyl alcohol–benzene–boehmite, adsorption is found not to be reversible with respect to temperature.

Similar chemisorption has been found with related alcohols and with adsorbents such as silica gel and titania gel.[41] Irreversible adsorption of nitrogenous bases by these gels also occurs.[48] The discovery of chemisorption by oxide gels at room temperature may have consequences in chromatography (Chapter 16).

When chemisorption occurs, it must normally be supposed that preferential physical adsorption of one component also occurs on top of the chemisorbed layer. The extent of this physical adsorption can be calculated if the extent of chemisorption is first determined from other experiments (e.g. adsorption from the vapour phase). Equation (3.2) can then be modified to give:

$$\frac{n_0 \Delta x}{m} = [(n_1^s)_{\text{chem}} + (n_1^s)_{\text{phys}}](1-x) - n_2^s x, \tag{4.35}$$

where $(n_1^s)_{\text{chem}}$ and $(n_1^s)_{\text{phys}}$ represent the numbers of moles of component 1 chemically and physically adsorbed, respectively, by unit weight of solid. $(n_1^s)_{\text{chem}}$ is independent of x, but $(n_1^s)_{\text{phys}}$ varies with x in the usual way.

Alternatively, the solid can be exposed to component 1 until chemisorption is complete, and the modified adsorbent can be allowed to adsorb both components physically from mixtures. These two procedures give the same results (Fig. 4.25).[44] In calculating individual isotherms, monolayer values

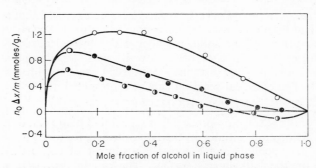

FIG. 4.25. Composite isotherm for physical adsorption on boehmite from mixtures of alcohol and benzene. Curves calculated after allowing for chemisorption on untreated boehmite; points, direct results for physical adsorption on previously "alkylated" boehmite: ○, methyl alcohol and benzene; ●, ethyl alcohol and benzene; ◑, n-butyl alcohol and benzene.[44] (Reproduced with permission from the *Journal of the Chemical Society*.)

must be determined for the modified adsorbent. It is seen that even the modified adsorbents show a strong preference for alcohols as compared with benzene (Fig. 4.25); "alkylated" boehmite still has two unchanged hydroxyl groups for every chemisorbed alcohol molecule, though the chemisorbed alkyl groups may prevent access to some of the remaining alkyl groups.

Very strong adsorption, which probably amounts to chemisorption, has been found when solutions of the lower fatty acids and alcohols are brought into contact with the carbon black, Spheron 6; physical adsorption also occurs on the chemisorbed layer. It is possible[49] that acetic acid reacts with "n-lactone" groups believed[50] to be present on the surface of this type of carbon black.

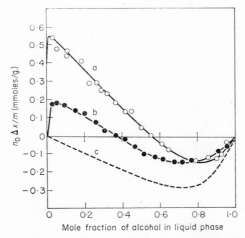

Fig. 4.26. Adsorption on Spheron 6 from mixtures of alcohols and benzene: (a) methyl alcohol and benzene, (b) ethyl alcohol and benzene, (c) methyl alcohol and benzene after correction for chemisorption.[26] (Reproduced with permission from "Proceedings of the Fourth Conference on Carbon", Pergamon Press.)

In adsorption from mixtures of methyl alcohol and benzene on Spheron 6 the composite isotherm is S-shaped (Fig. 4.26) but shows an unusual disproportion between the two parts, the first having a sharp peak, and the second a shallow trough; the peak is much higher than that in isotherms for mixtures of benzene with ethyl, propyl, and butyl alcohols.[26] Although the peak is not so high as to require multilayer adsorption to be postulated, the usual analysis would give individual isotherms showing both a maximum and a minimum.

Irreversible adsorption of methyl alcohol from the vapour phase was observed, and an approximate analysis of the composite isotherm was made by assuming that chemisorption also occurred during adsorption from the liquid phase. The residual physical adsorption involved strong preferential adsorption of benzene. The chemisorption is attributable to oxygen complexes on the carbon black. Once these are masked by chemisorption, the surface loses its polarity. It then behaves like Graphon, which has no oxygen complexes on the surface, in adsorbing benzene preferentially to alcohol. This case is different from those of the oxide gels, which retain some polar centres after chemisorption of alcohol has taken place. Chemisorption on the carbon black leaves accessible no significant number of polar centres.

Application of the Schay–Nagy analysis can often be made to systems in which chemisorption takes place, as the isotherms frequently have sub-

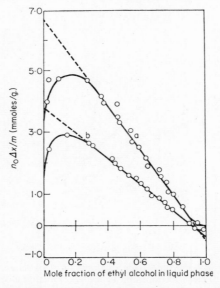

FIG. 4.27. Composite isotherms for adsorption on silica gel from mixtures of ethyl alcohol and benzene at 25°c, showing effect of chemisorption: (a) adsorption on untreated silica gel, (b) adsorption on silica gel pretreated with ethyl alcohol.[31] (Reproduced with permission from *Periodica Polytechnica*.)

stantial linear sections. In this case, the calculated amount of adsorbate includes both chemisorbed and physically adsorbed material. The amount chemisorbed may be very considerable, as is illustrated in Fig. 4.27. Ethyl alcohol is both chemisorbed and physically adsorbed from mixtures with benzene by silica gel.[31] If the gel is pre-treated with alcohol, only physical adsorption occurs from solution, and the extrapolated value for physical adsorption is little more than half of the total adsorption.

A further complex case is found in adsorption on carbon black from solutions of bromine in carbon tetrachloride.[51] The composite isotherms

(Fig. 4.28) have maxima too high to correspond to monolayer adsorption of bromine. It seems likely that chemisorption as well as physical adsorption of bromine occurs on Spheron 6. There is evidence that bromine is chemisorbed from solutions in carbon tetrachloride by coal chars.[52] On Graphon, evidence from adsorption of bromine vapour suggests that intercalation between graphite layers may occur in addition to physical adsorption on the external surfaces of graphitic lamellae.

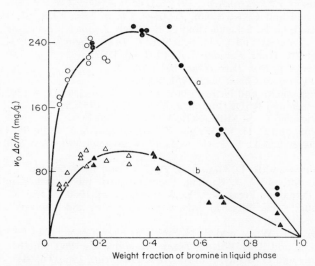

FIG. 4.28. Composite isotherms for adsorption (a) on Spheron 6, and (b) on Graphon from mixtures of bromine and carbon tetrachloride at 20°C: open symbols, adsorption from solution; closed symbols, adsorption from mixed vapours.

Adsorption of bromine from aqueous solutions is more complex. There is a tendency for conversion to hydrogen bromide to take place, but if the charcoal has been evacuated to a high temperature (750–1200°C) immediately before contact with the solution, the main process tends to be chemisorption of bromine.[53] The corresponding adsorption of chlorine from solution is discussed in Chapter 6, and of iodine in Chapter 7.

REFERENCES

1. A. M. Williams, *Medd K. Vetenskapsakad. Nobelinst.*, 1913, **2**, No. 27.
2. A. C. Zettlemoyer, Mattiello Award Lecture, 1958.
3. J. J. Kipling, *J. Chem. Soc.*, 1952, 2858.
4. S. Fénéant-Eymard, *Mém. Services chim. État.*, 1952, **37**, 297.
5. B. Gustafson, *Z. phys. Chem.*, 1916, **91**, 385.
6. F. G. Tryhorn and W. F. Wyatt, *Trans. Faraday Soc.*, 1925, **21**, 399; 1926, **22**, 134, 139; 1928, **24**, 36.

68 ADSORPTION FROM SOLUTIONS OF NON-ELECTROLYTES

re

7. J. J. Kipling and D. A. Tester, *J. Chem. Soc.*, 1952, 4123.
8. A. Blackburn and J. J. Kipling, *J. Chem. Soc.*, 1954, 3819.
9. W. B. Innes and H. H. Rowley, *J. Phys. Chem.*, 1947, **51**, 1154, 1172.
10. A. Blackburn and J. J. Kipling, *J. Chem. Soc.*, 1955, 1493.
11. D. C. Jones and L. Outridge, *J. Chem. Soc.*, 1930, 1574.
12. A. M. Bakr and J. W. McBain, *J. Amer. Chem. Soc.*, 1924, **46**, 2718.
13. W. Ostwald and R. de Izaguirre, *Kolloid-Z.*, 1922, **30**, 279.
14. D. Schmidt-Walter, *Kolloid Z.*, 1914, **14**, 242.
15. B. Gustafson, *Z. Elektrochem.*, 1915, **21**, 459; *Z. phys. Chem.*, 1916, **91**, 385.
16. F. E. Bartell and C. K. Sloan, *J. Amer. Chem. Soc.*, 1929, **51**, 1637, 1643; F. E. Bartell, G. H. Scheffler, and C. K. Sloan, *J. Amer. Chem. Soc.*, 1931, **53**, 2501; F. E. Bartell and G. H. Scheffler, *J. Amer. Chem. Soc.*, 1931, **53**, 2507.
17. D. H. Everett, *Trans. Faraday Soc.*, 1964, **60**, 1803.
18. A. Klinkenberg, *Rec. Trav. chim.*, 1959, **78**, 83.
19. G. A. H. Elton, *J. Chem. Soc.*, 1951, 2958.
20. E. A. Guggenheim and N. K. Adam, *Proc. Roy. Soc.*, 1933, A **139**, 218.
21. R. S. Hansen and R. D. Hansen, *J. Colloid Sci.*, 1954, **9**, 1.
22. D. C. Jones and G. S. Mill, *J. Chem. Soc.*, 1957, 213.
23. J. J. Kipling and E. H. M. Wright, *Trans. Faraday Soc.*, 1959, **55**, 1185.
24. A. V. Kiselev and L. F. Pavlova, *Kinetics and Catalysis*, (*U.S.S.R.*), 1961, **2**, 542.
25. R. S. Hansen and R. D. Hansen, *J. Phys. Chem.*, 1955, **59**, 496.
26. C. G. Gasser and J. J. Kipling, Proceedings of the Fourth Conference on Carbon, Pergamon Press, New York and London, 1960, p. 55.
27. S. Brunauer, "The Physical Adsorption of Gases and Vapours", Princeton and Oxford, 1944.
28. C. Pierce, J. W. Wiley, and R. N. Smith, *J. Phys. Chem.*, 1949, **53**, 669.
29. J. J. Kipling and R. B. Wilson, *Trans. Faraday Soc.*, 1960, **56**, 562.
30. L. G. Nagy and G. Schay, *Magyar Kém. Folyóirat.*, 1960, **66**, 31.
31. G. Schay, L. G. Nagy, and T. Szekrenyesy, *Periodica Polytech.*, 1960, **4**, 95.
32. G. Schay and L. G. Nagy, *J. Chim. phys.*, 1961, 149.
33. P. V. Cornford, J. J. Kipling, and E. H. M. Wright, *Trans. Faraday Soc.*, 1962, **58**, 74.
34. C. L. Lloyd and B. L. Harris, *J. Phys. Chem.*, 1954, **58**, 899.
35. G. Delmas and D. Patterson, *Off. Dig. Fed. Paint Varn. Prod. Cl.*, 1959, **31**, 1129.
36. M. R. Cines and F. N. Ruehlen, *J. Phys. Chem.*, 1953, **57**, 710.
37. J. J. Kipling, *J. Colloid Sci.*, 1963, **18**, 502.
38. M. Polanyi, *Verhand. deut. phys. Ges.*, 1916, **18**, 55.
39. M. Polanyi, *Z. Physik*, 1920, **2**, 117.
40. R. S. Hansen and W. V. Fackler, *J. Phys. Chem.*, 1953, **57**, 634.
41. J. J. Kipling and D. B. Peakall, *J. Chem. Soc.*, 1957, 4054.
42. M. R. A. Rao, *J. Indian Chem. Soc.*, 1935, **12**, 371.
43. T. E. Groves, S. T. Bowden, and W. J. Jones, *Rec. Trav. chim.*, 1947, **66**, 645.
44. J. J. Kipling and D. B. Peakall, *J. Chem. Soc.*, 1956, 4828.
45. J. J. Kipling, "Proceedings of the Second International Congress of Surface Activity", Butterworths, London, 1957, Vol. III, p. 462.
46. J. J. Kipling and D. B. Peakall, *J. Chem. Soc.*, 1957, 834.

47. J. J. Kipling and D. B. Peakall, "Chemisorption", (W. E. Garner, ed.), Butterworths, London, 1957, p. 59.
48. J. J. Kipling and D. B. Peakall, *J. Chem. Soc.*, 1958, 184.
49. J. J. Kipling and E. H. M. Wright, *J. Phys. Chem.*, 1963, **67**, 1789.
50. V. A. Garten, D. E. Weiss, and J. B. Willis, *Austral. J. Chem.*, 1957, **10**, 295.
51. P. V. Shooter, unpublished results.
52. J. D. Brooks and T. McL. Spotswood, "Proceedings of the Fifth Conference on Carbon", Pergamon Press, Oxford, 1962, p. 416.
53. B. R. Puri, "Proceedings of the Fifth Conference on Carbon", Pergamon Press, Oxford, 1962, p. 165.

CHAPTER 5

Adsorption from Partially
Miscible Liquids

THE S-SHAPED ISOTHERM

Isotherms for adsorption from binary systems of this type are usually of
the shape shown in Fig. 5.1 for the adsorption by silica gel from solutions
of methyl alcohol in heptane.[1] These isotherms have been described as
S-shaped, but must be distinguished from the S-shaped isotherms found for
adsorption from many pairs of completely miscible liquids. The
characteristic is that the extent of adsorption increases as the solubility limit
is approached, and appears to become asymptotic to a line parallel to the
adsorption axis.

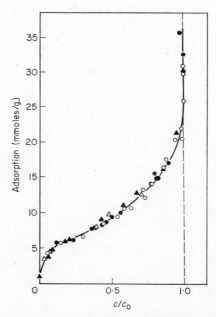

FIG. 5.1. Adsorption by silica gel from solutions of methyl alcohol in heptane[1]: circles,
0°c; triangles, 20°c; open symbols, adsorption; closed symbols, desorption. (Reproduced
from *Doklady Akad. Nauk S.S.S.R.*)

It is very difficult to determine the isotherm precisely at concentrations close to the solubility limit, but the trend suggests such a rapid rise that the adsorbed layer must almost certainly be considered to become multimolecular. A general suggestion of this kind was made many years ago with reference to some isotherms of rather unsatisfactory shape.[2] More recently it has been suggested that the sharp rise in the isotherm is due to incipient phase separation under the influence of the solid surface.[3]

Variations in the Shape of the Isotherm

On free surfaces, the shape of the isotherm shows little variation. For some homologous series, indeed, the isotherms are coincident if plotted in terms of activity or reduced concentration of the solute (Fig. 5.2).[4] For porous

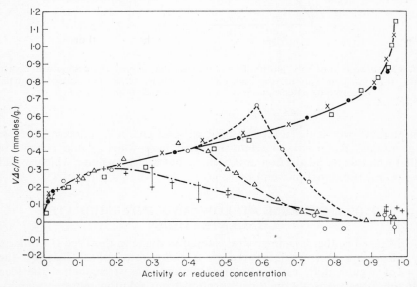

FIG. 5.2. Adsorption by graphite from a homologous series of fatty acids from aqueous solution[4]: +, acetic acid; △, propionic acid; ○, n-butyric acid; ✕, n-valeric acid ;●, n-caproic acid; □, n-heptylic acid. (Reproduced with the permission of the American Chemical Society from the *Journal of Physical Chemistry*.)

solids, however, the size of the pores may set a limit to the number of adsorbed layers which can be formed. Thus when butyl alcohol is adsorbed from water by a charcoal known from other evidence to have wide pores, the isotherm is fairly similar to that for a non-porous carbon (Fig. 5.3a). If the pores are narrow, however, the rise over the latter part of the concentration range, indicative of multilayer formation, does not take place (Fig. 5.3b).[1] The same phenomenon has been found in adsorption of phenol from aqueous solutions by various carbons.[3] Even with porous charcoals, there may be a

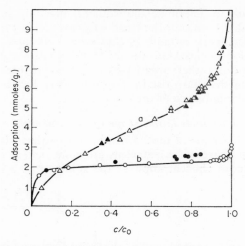

FIG. 5.3. Adsorption of butyl alcohol from aqueous solution on: (a) a carbon with coarse pores, (b) a carbon with narrow pores. Open symbols, adsorption; closed symbols, desorption.[1] (Reproduced from *Doklady Akad. Nauk. S.S.S.R.*)

considerable rise at the solubility limit as most charcoals contain pores of large radius. These have a relatively small area which would contribute little to multilayer adsorption at a surface, but a large volume which would take up the separated phase in bulk.

THE ISOTHERM AT LOW AND INTERMEDIATE CONCENTRATIONS

The S-shaped isotherm is, in general, similar in shape to the Type II isotherm for adsorption of a single vapour by a non-porous solid.[5] This, on the basis of argument by rather doubtful analogy, has strengthened the idea that the S-shaped isotherm for adsorption from solution represents multilayer adsorption at high relative concentrations.

Further, it was found that the equation put forward by Brunauer, Emmett, and Teller to describe the above type of vapour isotherm[5] would also fit the isotherms for adsorption from solution if p/p_0 were replaced by c/c_0. Figure 5.4 shows this for adsorption by an artificial graphite from aqueous solutions of n-valeric and n-caproic acids.[3] Similar curves are found for several other systems. There is a marked departure at high concentrations, but no more than is found at high relative pressures in the use of this equation for isotherms for adsorption of vapours.

It is well known that although a particular equation may fit a curve, this is not necessarily of any physical significance. No attempt has been made to justify the use of the modified B.E.T. equation in adsorption from solution

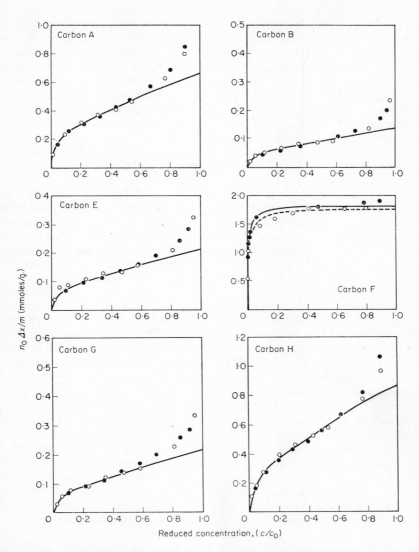

FIG. 5.4. Adsorption by carbons from aqueous solutions of n-valeric and n-caproic acids: ○ and – – – –, n-valeric acid; ● and ————, n-caproic acid; points, experimental; curves according to "B.E.T." equation. A, B, and E, artificial graphite; F, sugar charcoal (porous); H, carbon black.[3] (Reproduced with permission of the American Chemical Society from the *Journal of Physical Chemistry*.)

on theoretical grounds. It was found empirically that the constants did not have values corresponding to their significance in the original derivation (monolayer value of adsorbate, heat of adsorption, or, in the extended form of the equation, the limiting number of layers formed by the adsorbate).[3] It has been specifically stated that any agreement between surface areas based on such "monolayer values" and those calculated from adsorption of nitrogen is fortuitous.[6] The major limitation of the attempt to use the equation for adsorption from solution must be that, in general, both components of the solution are adsorbed over most of the concentration range, and especially at low concentrations.

It has been suggested that if the solute is very much more strongly adsorbed than the solvent, the "knee" in the isotherm might represent the formation of a complete monolayer of the solute on the solid surface, i.e. the solute is so strongly adsorbed that its behaviour alone essentially determines the shape of the composite isotherm. (In this respect, the composite isotherm can, for most of the systems being discussed, be taken as approximating to the isotherm for the adsorption of the solute itself, as the solute concentrations are low—cf. equation [3.9].) If this were the case, the modified B.E.T. equation could be used as a standard method for evaluating the extent of adsorption at the "knee". Some evidence is available from a comparison of the adsorption of triethylamine from aqueous solution and from the vapour phase, the isotherms[7] being plotted on the same scale of p/p_0 or c/c_0. The results are equivocal. For a carbon black the isotherms were identical up to a relative concentration of 0·6, but vapour adsorption then increased more than adsorption from solution until saturation was reached (Fig. 5.5(a)). On a charcoal with coarse pores, however, adsorption from solution was less at all stages except at saturation, implying that adsorption of the solvent was still significant (Fig. 5.5(b)).

For the adsorption of n-octane, n-decane, and n-dodecane from methanol by Graphon, the "knee" does correspond roughly to the value required for a monolayer of hydrocarbon, if the hydrocarbon molecules are oriented with the major axis parallel to the surface.[8] For adsorption on Spheron 6 and on a sample of graphite, however, the adsorption at the knee is lower, although the specific surface areas of the adsorbents are higher than that of Graphon (Fig. 5.6). This is presumably due to adsorption of the polar methanol molecules on oxygen sites on the surfaces of Spheron 6 and the graphite sample, i.e. the surfaces are regarded as being heterogeneous. Competition from the solvent seems unlikely to explain the situation shown in Fig. 5.7 for adsorption on Graphon from aqueous solutions of the lower fatty acids.[4] The extent of adsorption at the "knee" is too low to correspond to a complete monolayer of fatty acid (except possibly for the highest member), but water is unlikely to compete for the surface, which is known to be hydrophobic. Similar isotherms, moreover, are found with Spheron 6, and for adsorption of the lower alcohols from aqueous solution on both Graphon (Fig. 5.8) and Spheron 6.

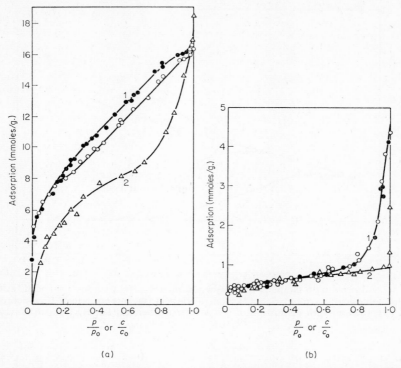

FIG. 5.5. Adsorption of triethylamine (1) from the vapour phase, (2) from aqueous solution, on (a) carbon black, (b) a porous carbon: open symbols, adsorption; closed symbols, desorption.[7] (Reproduced from *Doklady Akad. Nauk. S.S.S.R.*

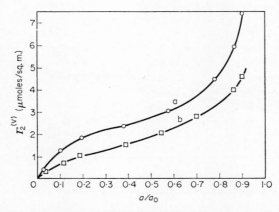

FIG. 5.6. Adsorption of n-decane from methyl alcohol[8] by (a) Graphon, (b) Spheron 6. (Reproduced with permission of the American Chemical Society from the *Journal of Physical Chemistry*.)

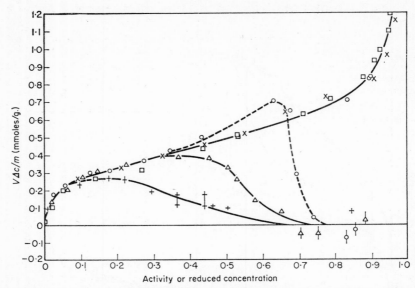

FIG. 5.7. Adsorption on Graphon from aqueous solutions of fatty acids[4]: +, acetic acid; △, propionic acid; ○, n-butyric acid; ✕, n-valeric acid; ●, n-caproic acid; □, n-heptylic acid. (Reproduced with permission of the American Chemical Society from the *Journal of Physical Chemistry*.)

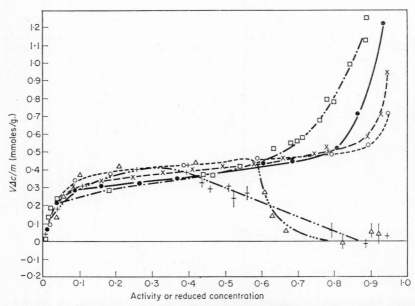

FIG. 5.8. Adsorption on Graphon from aqueous solutions of the lower alcohols[4]: +, ethanol; △, n-propanol; ○, n-butanol; ✕, n-pentanol; ●, n-hexanol; □, n-heptanol. (Reproduced with permission of the American Chemical Society from the *Journal of Physical Chemistry*.)

The isotherms for the adsorption on Graphon are less congruent than those for fatty acids. For n-butanol and higher members, distinctive values of adsorption at the "knee" can be obtained from Fig. 5.8, which correspond approximately to those which would be required for complete monolayers of alcohol molecules oriented with the major axis parallel to the surface. This agreement is not found for adsorption on Spheron 6, on which competition from the solvent is likely to be important. For systems of this kind, a closer examination of the isotherm in the region of the "knee" is desirable.

For both adsorbents, the adsorption observed at the highest accessible concentrations is greater than corresponds to a monolayer of alcohol in the closest possible packing (with the molecules perpendicular to the surface). There is no doubt that, at such concentrations, multilayer adsorption takes place.

Fu, Hansen, and Bartell[9] commented on the significance of congruence in such systems in terms of the work required to remove a solute from solution as the pure solute, which depends on the absolute activity of the solute. Thus, if the adsorbent acted on the same functional group, congruency of adsorption isotherms in a homologous series when plotted in terms of reduced concentration (which is an approximation for absolute activity) could readily occur. The difficulty in applying this suggestion to the present case is to envisage what is the required functional group. As Graphon has no polar groups on the surface, it cannot be the hydroxyl group, because the isotherms for adsorption on Spheron 6 and Graphon are very similar. If, on the other hand, the alcohol molecules were oriented with the hydroxyl groups towards the solution and the methyl groups towards the solid surface, it would be expected that the competition with water would be more serious on Spheron 6 than on Graphon.

As multilayer adsorption seems likely to occur in these systems, the application of the potential theory (see Chapter 4) may be relevant. It has been applied to one rather special case, that of adsorption by Spheron 6 (1000°c) from aqueous solutions of butanol.[10] The composite isotherm could be fitted at the higher relative concentrations only if an additional assumption, special to the system, was made.

THE ISOTHERM AT HIGHER CONCENTRATIONS

At a concentration of *mixture* beyond the limit for true *solution*, two phases exist and one of these wets the solid preferentially. It thus seems reasonable to think in terms of incipient phase separation under the influence of the solid surface at concentrations just below the solubility limit.[3] A porous solid may take up one of two immiscible phases to fill its pore system. This is the closest approach, in adsorption from solution, to the capillary condensation which takes place in adsorption of vapours by porous solids.

The possibility that phase separation would occur in the pores of a porous solid below the critical solution temperature was envisaged by Jones and

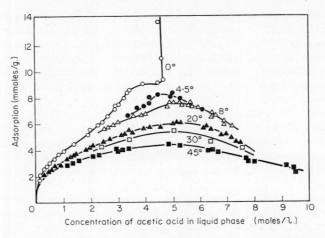

FIG. 5.9. Adsorption by silica gel from mixtures of acetic acid and heptane, above and below the critical solution temperature.[12] (Reprinted from *Doklady Akad. Nauk. S.S.S.R.*)

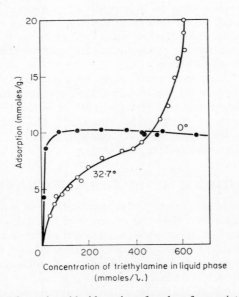

FIG. 5.10. Adsorption by a zinc chloride activated carbon from mixtures of triethylamine and water[7] above and below the lower critical solution temperature of 18°C. (Reprinted from *Doklady Akad. Nauk S.S.S.R.*)

Outridge.[11] Kiselev demonstrated this experimentally, for adsorption by silica gel from mixtures of acetic acid and heptane, by examining the shape of the isotherms above and below the critical solution temperature (Fig. 5.9).[12] Above the critical temperature, the isotherms had the normal shape for a completely miscible system, passing through a maximum. Below the critical temperature, the isotherm rose indefinitely at the solubility limit, corresponding to the filling of pores with a saturated solution of acetic acid in heptane.

In the above systems, the two liquids are completely miscible *above* the critical temperature. Adsorption has also been examined from systems completely miscible *below* the critical temperature. In this case there is a similar change in the opposite direction with respect to temperature, e.g. in the adsorption of triethylamine from aqueous solutions.[7] An extra feature is that the isotherms for different temperatures now intersect (Fig. 5.10). At low concentrations, adsorption decreases with rise in temperature, but at higher concentrations, it rises with temperature, as multimolecular adsorption or pore-filling by one phase replaces what is probably monomolecular adsorption.

Effect of Temperature

The effect of increase in temperature on adsorption from partially miscible liquids is generally to reduce selectivity. This can also be expressed in terms of the solvent becoming a "better" solvent for the solute at a higher tempera-

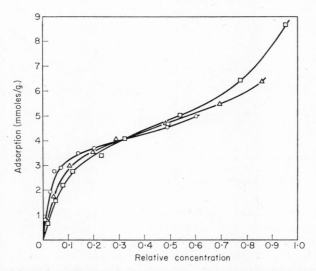

FIG. 5.11. Adsorption of n-butyl alcohol from aqueous solutions by graphite[13] at 0°c (○), 25°c (△), and 45°c (□). (Reproduced with permission of the American Chemical Society from the *Journal of Physical Chemistry*.)

ture, with corresponding reduction in adsorption of the solute. If, however, solubility has a negative temperature coefficient, adsorption may increase with rise in temperature. In the adsorption by graphite from aqueous solutions of n-butyl alcohol, adsorption decreases with increasing temperature at low relative concentrations; this is the normal effect, adsorption being an exothermic process. At higher relative concentrations, however, adsorption increases with temperature (Fig. 5.11), and this is attributed to the decrease in solubility which causes an increase in chemical potential at a given concentration, and consequently an increase in adsorption.[13]

Solid and Liquid Solutes

A further effect of lowering temperature is that the liquid solute may become solid. The isotherm then changes to that characteristic of the adsorption of solids from solution. This has been shown for the adsorption by silica gel from solutions of phenol in heptane (Fig. 5.12).[14] The isotherm is S-shaped,

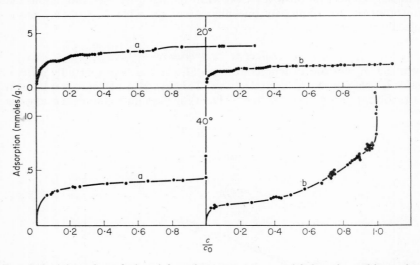

FIG. 5.12. Adsorption of phenol from heptane near to and below the melting point of phenol from (a) a fine-pored and (b) a coarse-pored silica gel.[14] Melting point of phenol: 40·9°C. (Reprinted from *Doklady Akad. Nauk S.S.S.R.*)

corresponding to multilayer formation, at 40°C, but reaches a limiting value, corresponding to monolayer formation, at 20°C. In the former case the volume adsorbed at saturation corresponds to the volume of pores in the solid; in the latter, the limiting adsorption corresponds to complete coverage of the surface area. For a gel with very narrow pores, the isotherms are almost identical except that the 40°C isotherm rises sharply at saturation whereas the 20°C isotherm does not.

The Second Section of the Isotherm

For partially miscible liquids, the composite isotherm should consist of two sections, one relating to a dilute solution of A in B, and the second to a dilute solution of B in A. Only rarely have both sections been examined for the same system.

Craig and Hansen[15] continued the investigations referred to above to give the adsorption of water from dilute solutions in alcohols and fatty acids. No change in concentration could be found with Graphon or graphite, which are hydrophobic. Adsorption of water would be negligible, and the consequent preferential adsorption of the alcohol or acid would not be measurable from a solution very highly concentrated in these components.

Adsorption of water by Spheron 6 was significant, and this accords with the presence of polar groups on the surface of this material. The composite isotherms were difficult to measure accurately, but appeared not to be sigmoid in character, i.e. not to rise sharply as the solubility limit was approached. This is most likely to be due to adsorption of water molecules at specific sites covering a part only of the surface, a process not likely to be conducive to the formation of multilayers of water.

Adsorption on Graphon has been examined at both ends of the concentration range from mixtures of formic acid and benzene.[16] Preferential adsorption of the component present in low concentration was observed in each case (Fig. 5.13), but there was no evidence of multilayer formation at either end of the miscibility range (Fig. 5.14). Multilayer formation does, however, occur in adsorption on Spheron 6 from solutions in carbon tetrachloride; see below.

CHEMISORPTION

As in adsorption from completely miscible liquids (Chapter 4), physical and chemical adsorption may occur together for some systems at room temperature. Figure 5.15 shows the isotherm for adsorption from solutions of formic acid in carbon tetrachloride onto Spheron 6. The curve appears to start from a finite value of adsorption instead of zero. Measurements of adsorption of formic acid from the vapour phase indicate the extent of chemisorption. An isotherm for physical adsorption only can then be calculated as described in Chapter 4, and the individual isotherms are obtained in the usual way.[17] These are shown in Fig. 5.16, which clearly indicates that multilayer formation occurs at remarkably low relative concentrations (x/x_0). There is no evidence for chemisorption on Graphon, and the preferential physical adsorption of formic acid is much less than on Spheron 6.

A special case of chemisorption occurs with charcoal and bromine water. (In some investigations, the aqueous solution has also contained potassium bromide; this, however, does not seem to have affected the chemisorption.) Freundlich[18] thought that unless the temperature were low, charcoal would

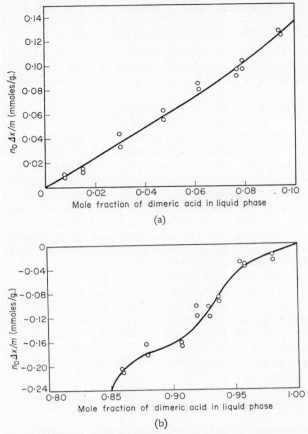

FIG. 5.13. Adsorption on Graphon from mixtures of formic acid and benzene: (a) at low concentrations of formic acid, (b) at high concentrations of formic acid.[16]

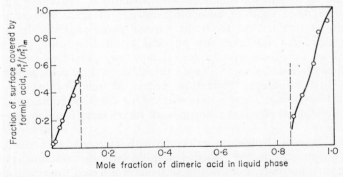

FIG. 5.14. Calculated surface coverage of Graphon in adsorption from mixtures of formic acid and benzene; broken lines show limits of miscibility.[16]

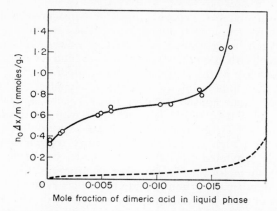

FIG. 5.15. Adsorption (composite isotherm) from formic acid–carbon tetrachloride solutions on Spheron 6 at 20°C. The broken line is the isotherm for adsorption on Graphon.[17] (Reproduced with permission of the American Chemical Society from the *Journal of Physical Chemistry*.)

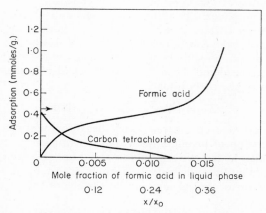

FIG. 5.16. Adsorption (individual isotherms) for physical adsorption from formic acid–carbon tetrachloride solutions on Spheron 6 at 20°C: x/x_0 shows the mole fraction of a given solution relative to that at the solubility limit. The arrow shows the monolayer value for physical adsorption of formic acid.[17] (Reproduced with permission of the American Chemical Society from the *Journal of Physical Chemistry*.)

not only adsorb bromine but would also assist the conversion of bromine water to hydrogen bromide and oxygen; he had more definite evidence of chemical reaction when charcoal was added to chlorine water (cf. Chapter 6). This view was supported by Kolthoff, who thought that the primary process was adsorption of bromine, even though he had found that hypobromous acid is very rapidly destroyed by charcoal.[19]

Puri has shown that two processes must be considered. Some charcoals bring about the reduction of bromine to hydrogen bromide, an equivalent quantity of oxygen being accounted for either as gaseous carbon dioxide or as oxygen chemisorbed on the charcoal surface (and recoverable as carbon dioxide by evacuation of the charcoal at high temperature).[20] This suggests the possibility that hypobromous acid might be adsorbed initially, thus making the process analogous to that occurring between charcoal and chlorine water (Chapter 6). These results are obtained with charcoals (e.g. sugar charcoal) having a high content of combined oxygen. If the charcoal is first evacuated to a high temperature to remove oxygen complexes, very little bromine is converted to hydrogen bromide, but large amounts are chemisorbed by the charcoal. The number of bromine atoms chemisorbed is equal to the number of carbon dioxide molecules evolved during evacuation in the form of carbon dioxide (the amount removed as carbon monoxide or steam being irrelevant).[21]

The first reaction predominates with sugar charcoal prepared at low temperatures, but takes place to decreasing extents as the temperature of evacuation of the charcoal is raised, becoming of little importance for charcoals evacuated at 700°c or above; the chemisorption of bromine by the charcoal becomes, correspondingly, increasingly important.

In adsorption on carbon black, Watson and Parkinson observed both physical and chemical adsorption of bromine.[22] Hydrogen bromide was shown to be produced, possibly by the mechanism suggested above. A further possibility to be considered, however, was direct displacement of hydrogen from the surface of the carbon black:

$$\geq C—H + Br_2 \rightarrow \geq C—Br + HBr.$$

This latter process was considered likely because hydrogen bromide was formed in adsorption of bromine from solution in carbon tetrachloride,† a solvent from which hydrolytic adsorption could not take place. Watson has further suggested that bromine may also form hydrogen bromide by oxidation of hydroquinone or alcohol groups on the surface of the carbon black.[23]

REFERENCES

1. O. M. Dzhigit, A. V. Kiselev, and K. G. Krasilnikov, *Doklady Akad. Nauk S.S.S.R.*, 1947, **58**, 413.
2. W. E. Garner, M. McKie, and B. C. J. G. Knight, *J. Phys. Chem.*, 1922, **31**, 641.
3. R. S. Hansen, Y. Fu, and F. E. Bartell, *J. Phys. Chem.*, 1949, **53**, 769.
4. R. S. Hansen and R. P. Craig, *J. Phys. Chem.*, 1954, **58**, 211.
5. S. Brunauer, "The Physical Adsorption of Gases and Vapours", Princeton and Oxford, 1944.
6. R. S. Hansen, R. D. Hansen, and R. P. Craig, *J. Phys. Chem.*, 1953, **57**, 215.

† Kolthoff thought that a solution of bromine in carbon tetrachloride attacked charcoal only slowly if at all.

7. A. V. Kiselev and V. V. Kulichenko, *Doklady Akad. Nauk S.S.S.R.*, 1952, **82,** 89.
8. R. S. Hansen and R. D. Hansen, *J. Phys. Chem.*, 1955, **59,** 496.
9. Y. Fu, R. S. Hansen, and F. E. Bartell, *J. Phys. Chem.*, 1948, **52,** 374.
10. R. S. Hansen and W. V. Fackler, *J. Phys. Chem.*, 1953, **57,** 634.
11. D. C. Jones and L. Outridge, *J. Chem. Soc.*, 1930, 1574.
12. K. G. Krasilnikov and A. V. Kiselev, *Doklady Akad. Nauk S.S.S.R.*, 1949, **69,** 817.
13. F. E. Bartell, T. L. Thomas, and Y. Fu, *J. Phys. Chem.*, 1951, **55,** 1456.
14. K. G. Krasilnikov and A. V. Kiselev, *Doklady Akad. Nauk S.S.S.R.*, 1948, **63,** 693.
15. R. P. Craig and R. S. Hansen, U.S. Atomic Energy Commission, Report ISC-309, Oak Ridge, Tennessee, 1953.
16. E. H. M. Wright, unpublished results.
17. J. J. Kipling and E. H. M. Wright, *J. Phys. Chem.*, 1963, **67,** 1789.
18. H. Freundlich, *Z. phys. Chem.*, 1907, **57,** 385.
19. I. M. Kolthoff, *Rec. Trav. chim.*, 1929, **48,** 291.
20. B. R. Puri, O. P. Mahajan, and D. D. Singh, *J. Indian Chem. Soc.*, 1961, **38,** 943.
21. B. R. Puri, N. K. Sandle, and O. P. Mahajan, *J. Chem. Soc.*, 1963, 4880.
22. J. W. Watson and D. Parkinson, *Ind. Eng. Chem.*, 1955, **47,** 1053.
23. J. W. Watson, Ph.D. Thesis, University of London, 1957.

CHAPTER **6**

Adsorption of Gases from Solution

NON-REACTING SYSTEMS

Little attention has been paid to this aspect of adsorption except in respect of gases which ionize in water (e.g. hydrogen chloride) and which behave as electrolytes. These are not considered here.

The adsorption of ethylene by charcoal at a given pressure is very much less if the charcoal is suspended in water than if it is dry. This is because, in the former case, the ethylene can only come into contact with the charcoal through the medium of its saturated solution in water. From the Bunsen coefficient, the concentration in solution can be calculated as a function of pressure in the gas phase. Alternatively, a "gas-pressure" corresponding to the concentration of the solution can be calculated. The adsorption isotherm, plotted against this equivalent "gas-pressure", can be fitted by a Freundlich equation. The pre-exponential factor is about one-tenth of that for adsorption directly from the gas phase, whereas the exponential factor is about 1·5 times as large as that for direct adsorption, cf. Fig. 6.1. This shows that there is competition between the ethylene and water for the solid surface when adsorption takes place from solution.[1]

In general, it may be expected that adsorption of most gases from solution will follow this pattern. The extent of adsorption is likely to be very small unless the gas reacts with the solid surface. Consequently systems of this kind are neither easy to investigate experimentally, nor of much practical interest.

ADSORPTION WITH REACTION

Adsorption from Chlorine Water

A more complex process occurs in the adsorption of chlorine from aqueous solution by water. This is essentially different from the previous system because chemical reaction takes place between the solute and solvent. The possibility therefore exists of extensive adsorption appearing to take place, but the adsorbed species being actually a reaction product.

The phenomenon is of technological interest in the dechlorination of water supplies and has been reviewed by Magee.[2] It is usually carried out as a column process. The most important step now appears to be adsorption of hypochlorous acid. This subsequently decomposes, on the surface, to give hydrochloric acid and nascent oxygen. The former is released into solution.

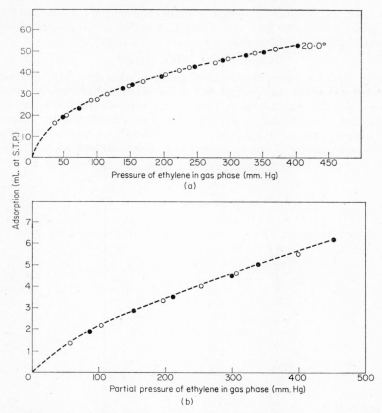

FIG. 6.1. Adsorption of ethylene on charcoal[1] (a) from the gas phase, (b) from aqueous solution, at 20°C. (Reproduced with permission from *Acta Chimica Acad. Sci. Hung.*)

The latter reacts with the charcoal to form a surface complex which may slowly decompose to give carbon monoxide and carbon dioxide. The breakdown is likely to be incomplete, and the retention of some oxygen lowers the efficiency of the charcoal in further dechlorination. The nature of the transient complexes is not clear. The existence of the permanent complexes, however, is demonstrated by heating the charcoal to 400°C, when the two oxides of carbon are evolved in large quantities and the efficiency of the charcoal in dechlorination is restored.

Using static (i.e. batch) conditions, Puri confirmed the formation of hydrochloric acid, but was unable to detect the evolution of oxides of carbon during the adsorption process.[3] He did, however, confirm the chemisorption of oxygen by the charcoal, the quantity recoverable being almost equivalent to the hydrochloric acid formed in the early stages of the process. In the later stages, the amount of oxygen chemisorbed reached a constant value, but hydrochloric acid was still formed, together with a corresponding quantity of chloric acid.[4] It thus appears that the reaction

$$3Cl_2 + 3H_2O \rightarrow 5HCl + HClO_3$$

occurs in the later stages, the charcoal acting as a catalyst. The observation of this latter reaction may depend on the use of solutions more concentrated (2660 p.p.m.) than those used in Magee's experiments (10–100 p.p.m.).

No evidence was found for the chemisorption of chlorine by charcoal.[5] In this respect chlorine in aqueous solution differs from bromine in aqueous solution (see Chapter 5) and in non-aqueous solutions (see below).

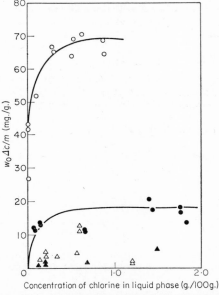

FIG. 6.2. Adsorption by carbon blacks from solutions of chlorine in carbon tetrachloride at 20°C: ○, Spheron 6; ●, Spheron 6 (1000°C); △, Spheron 6 (1400°C); ▲, Graphon.[6]

Adsorption of Chlorine from Non-aqueous Solutions

The adsorption of chlorine from non-aqueous solutions should be simpler. There are few simple organic solvents, however, with which chlorine does not react at an appreciable rate at room temperature. The most useful for investigating adsorption is carbon tetrachloride.

The adsorption of chlorine from carbon tetrachloride on a series of related carbon blacks is shown[6] in Fig. 6.2. On Graphon and Spheron 6 (1400°C),† adsorption is too small to be measured accurately. As these

† Spheron 6 (1400°C) refers to the carbon formed by heating Spheron 6 to 1400°C in an inert atmosphere.

carbons have practically no surface complexes, there are no centres for chemisorption, and it is probable that only physical adsorption takes place. In this process, the important feature is the volatility of both chlorine and the solvent, as each presents chlorine atoms to the surface of the absorbent.

Extensive adsorption occurs on Spheron 6 (1000°c), which retains some hydrogen on the surface, and on unheated Spheron 6, the surface of which is probably covered with oxygen and hydrogen complexes. The majority of the chlorine is probably chemisorbed, but as the isotherm for Spheron 6 shows dependence on concentration, some physical adsorption may also occur. Chemisorption in this case could involve addition of chlorine to double bonds or displacement of hydrogen with simultaneous formation of hydrogen chloride. If the latter process occurred, the points in Fig. 6.2 would represent total removal of chlorine from solution, and only a part of this amount would be adsorbed, the remainder being returned to solution as hydrogen chloride.

REFERENCES

1. T. Erdey-Grúz and F. Nagy, *Acta Chim. Acad. Sci. Hung.*, 1957, **12,** 101.
2. V. Magee, Ph.D. Thesis, University of London, 1955.
3. B. R. Puri, D. D. Singh, J. Chander, and L. R. Sharma, *J. Indian Chem. Soc.*, 1958, **35,** 181.
4. B. R. Puri, O. P. Mahajan, and D. D. Singh, *J. Indian Chem. Soc.*, 1960, **37,** 171.
5. B. R. Puri, "Proceedings of the Fifth Conference on Carbon", Pergamon Press, Oxford, 1962, p. 165.
6. P. V. Shooter, unpublished results.

CHAPTER 7

Adsorption of Solids from Solution

INTRODUCTION

In the practical use of adsorption, solutions of solids are encountered much more frequently than the systems described in previous chapters. In this chapter the adsorption of solids of relatively low molecular weight is considered. A special case of adsorption of solids from solution is that of polymeric materials, in which interest has increased considerably recently. These systems present a number of special features not found when smaller molecules are involved. They are therefore treated separately, in Chapter 8.

A number of general features emerge from a study of the simpler solutes. In addition, most of the types of system have considerable inherent interest, and detailed discussion of individual systems is consequently required. As the systems can be studied only over a restricted range of concentration, the terms "solute" and "solvent" are appropriate for the two components of a binary mixture.

In the academic study of adsorption from this type of system, the following aspects have emerged as being of greatest interest and importance:

(i) the shape of the adsorption isotherm and the possibility of fitting it with an appropriate equation,
(ii) the significance of the adsorption limit or the plateau found in most isotherms,
(iii) the extent to which the solvent is adsorbed,
(iv) a consideration of whether adsorption is confined to a single molecular layer or extends over several layers,
(v) the orientation of the adsorbed molecules,
(vi) the existence of both physical and chemical adsorption.

The significance of these topics is first illustrated, and several important systems are then described in some detail.

Early Experiments

Much of the original work on adsorption from solutions of solids (as on adsorption from solution generally) is described by Freundlich,[1] who carried out a great deal of the experimental work. In considering adsorption from dilute solutions, he made no distinction between solid and liquid solutes. Indeed, it is not yet clear how useful it will prove ultimately to make such a distinction. At the time of writing, however, a number of differences are appearing which suggest that separate treatments may be useful at this stage of the development of the subject.

The early experiments were almost entirely confined to aqueous solutions, and charcoal of various kinds was almost universally used as the adsorbent. Adsorption from aqueous solutions of the lower fatty acids and alcohols, picric, succinic, and benzoic acids, phenol, acetone, and bromine could be represented very well by the so-called Freundlich equation (Chapter 3). The theoretical importance of this result is not yet clear, but it may be significant that the range of concentration studied was very small; the highest value was frequently about 0·1 M, which corresponds to a mole fraction of about 0·002 for aqueous solutions.

Freundlich did, however, investigate the influence of the solvent on adsorption; for example, he found that adsorption of benzoic acid by charcoal was, for the same concentrations, ten times less from solution in diethyl ether than from aqueous solution.[1] He related this to the interfacial tension given by the solvent in question against the adsorbent, a low interfacial tension being accompanied by low adsorption of the solute. On the other hand, for solvents of approximately the same interfacial tension adsorption is weaker, the greater the solubility of the solute. This was deduced from Polanyi's theory of adsorption and was shown experimentally by Lundelius in adsorption on charcoal at 19–20°c from solutions of iodine in carbon disulphide, chloroform, and carbon tetrachloride.[2] The concentrations of the three solutions needed to effect a given degree of adsorption, x/m, were in the ratio 4·5 : 2 : 1, whereas the corresponding solubilities at 14·5°c were in the ratio 4·8 : 1·8 : 1. The results were expressed in terms of the Freundlich equation (3.1). The value of the exponential term, $1/n$, was found to be the same for all three systems.

THE ADSORPTION ISOTHERM AND EQUATIONS FOR THE ISOTHERM

The commonest shape of isotherm for adsorption of a solid from solution is shown in Fig. 7.1. Many other shapes have been recorded, but each is found for a few systems only. A useful classification of such isotherms is

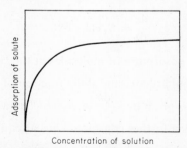

Fig. 7.1. Typical shape of isotherm for adsorption of a solid from solution.

discussed below (p. 129). The isotherm of the shape shown in Fig. 7.1 can usually be fitted by a Freundlich equation at low concentrations. If adsorption is examined up to the solubility limit, however, it is usually found that the curve becomes asymptotic to a limiting value of adsorption. It cannot, therefore, be fitted accurately by a Freundlich equation over the whole of the available range of concentration.

Although Schmidt[3] produced modified equations which took into account a limiting value for adsorption, attention has mainly been focused on an equation having the form of the Langmuir equation:

$$\frac{x}{m} = \frac{kKx}{1+kx}. \tag{7.1}$$

This attention probably resulted from the similarity of shape between the type of isotherm under discussion and the Type I isotherm found in adsorption of vapours by solids. As the latter was fitted satisfactorily by the Langmuir equation in its original form (with pressure terms), a modification including concentration terms was expected to apply to adsorption from solution.

The objections to this alteration of the Langmuir equation have been stressed in Chapter 4. Despite its inapplicability to adsorption from completely miscible liquids, however, it often fits the isotherm obtained for adsorption of solids from dilute solutions.

No complete explanation of this applicability of the equation has been given. A treatment has been proposed for adsorption at the liquid–vapour interface which should be applicable to the liquid–solid interface. Its basis is kinetic, as in the determination of the original Langmuir equation for adsorption of single gases. The kinetic treatment is applied to adsorption and desorption of the solute only. Thus, if the treatment is valid, it is implied that equilibrium is governed essentially by the way in which the solute is adsorbed, rather than by competition between solute and solvent for the interface. It may therefore be expected not to have universal validity. The derivation is based on work by Guastalla[4] and Davies,[5] and as presented here is modified from the account given by Davies and Rideal.[6]

The rate of adsorption of the solute, through a stagnant layer adjacent to the interface, is

$$\frac{\mathrm{d}n}{\mathrm{d}t} = B_1 c(1-\theta),$$

where $\frac{\mathrm{d}n}{\mathrm{d}t}$ is the rate of adsorption of molecules at the surface, c is the molar concentration in the liquid phase, and θ the fraction of the surface covered. The rate of desorption is given by $-\frac{\mathrm{d}n}{\mathrm{d}t} = B_2 n\,\mathrm{e}^{-W/kT}$ where W is the energy of adsorption and B_2 is a further constant. Thus at equilibrium

$$B_2 n\,\mathrm{e}^{-W/kT} = B_1 c(1-\theta),$$

and if

$$n = n_0 \theta,$$

$$n = \frac{B_1 c}{B_2 e^{-W/kT} + B_1 \dfrac{c}{n_0}}$$

$$= \frac{\dfrac{B_1 c}{B_2} e^{+W/kT}}{1 + \dfrac{B_1 c}{B_2 n_0} e^{W/kT}}$$

$$\therefore \quad n = \frac{n_0 \dfrac{B_1 c}{B_3} e^{W/kT}}{1 + \dfrac{B_1 c}{B_3} e^{W/kT}}, \qquad (7.2)$$

where $B_3 = B_2 n_0$; the terms n and n_0 are expressed in molecules per unit area, n_0 being the number present in a complete monolayer of solute. This equation has the form of the Langmuir equation provided that W is independent of θ. This is equivalent to Langmuir's postulate of an energetically homogeneous surface, together with the assumption that there is no lateral interaction between adsorbed molecules. The occurrence of such interaction would result in variations of W with θ. A typical case of compliance with this equation is given by the adsorption of n-cetylamine and of dimethyl-n-cetylamine from nujol by metal surfaces (mild steel, copper, aluminium, and nickel).[7]

The essential features of the Freundlich and Langmuir equations in the fitting of curves have been pointed out by Jowett.[8] The Freundlich equation has no asymptote, and the slope at the origin is infinite. The Langmuir equation gives an asymptote, the value of which is related to the slope at the origin. These characteristics arise because each equation involves two constants. Jowett considered that a more satisfactory fitting of experimental points could be obtained with three constants, and proposed the semi-empirical equation:

$$\frac{x}{m} = A - (A - a) e^{-Bc} \qquad (7.3)$$

in which A, a, and B are constants; A gives the value of the asymptote. This equation allows for the possibility that chemisorption as well as physical adsorption may occur, so that the curve may not pass through the origin. It was found to apply to the adsorption of m-cresol from aqueous solutions by coal.

In general, of course, a three-constant equation is likely to lead to better curve-fitting than is a two-constant equation. It remains to be seen whether

4*

the theoretical significance can be established for such an empirical relation, a situation which has not yet been achieved for either the Freundlich or the Langmuir equation as used to describe adsorption from solution.

In many systems the adsorption of the solvent cannot be neglected. Even in these cases, the isotherm can sometimes be fitted by this form of the Langmuir equation, but it cannot then be expected that the constants will have the significance attributed to them above.

Adsorption of the Solvent

Equation (3.2), relating the adsorption of the individual components to the value plotted for the composite isotherm,

$$n_0 \Delta x/m = n_1^s(1-x) - n_2^s x,$$

was derived in general terms and applies strictly to all systems. In principle, therefore, the adsorption of the solvent must be considered. For systems in which the adsorbed layer is thought to be confined to one molecular layer, the values of n_1^s and n_2^s can be calculated by using equation (4.20):

$$\frac{n_1^s}{(n_1^s)_m} + \frac{n_2^s}{(n_2^s)_m} = 1.$$

For the solid solute, $(n_1^s)_m$ cannot usually be obtained from measurements of the adsorption of the vapour, and the value is therefore usually calculated either from data for the crystal or from an assumed packing of the molecules, drawn to scale on the basis of accepted bond lengths and bond angles. As few molecules are symmetrical, the latter procedure normally involves a judgement as to the most likely orientation of the adsorbed molecules. There is no certain way of establishing this, and the reliability of any assumed orientation depends on the extent of supporting circumstantial evidence.

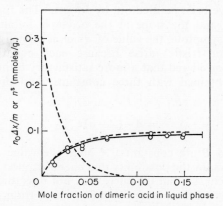

FIG. 7.2. Individual isotherms (broken lines) for adsorption on Graphon from solutions of lauric acid in carbon tetrachloride. The full line is the composite isotherm.[9] (Reproduced with permission from the *Journal of the Chemical Society*.)

Examples of calculated individual isotherms for adsorption of the solvent are shown in Figs. 7.2 and 7.3. One is for an aqueous and the other for a non-aqueous system.[9] In both of these cases the adsorption of the solvent is very high for dilute solutions. For these systems it is particularly high in relation to that of the solute, because the solvent molecules are considerably

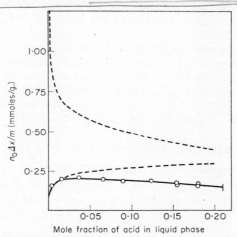

FIG. 7.3. Individual isotherms (broken lines) for adsorption on Graphon from aqueous solutions of malonic acid. The full line is the composite isotherm.[9] (Reproduced with permission from the *Journal of the Chemical Society*.)

smaller than those of the solute and a greater number of moles is therefore required to cover a given area. (The justification for considering the adsorbate in these two systems to be confined to a single molecular layer is given in later sections [pp. 97, 99].)

Adsorption of the solvent is often not considered in this type of system. Although it is rarely important for its own sake, its neglect may frequently lead to a faulty interpretation of the adsorption of the solute.

Adsorption of the Solute

The adsorption of the solute can also be calculated from equations (3.2) and (4.20). In Fig. 7.2, it is seen that the individual isotherm for adsorption of the solute lies close to the composite isotherm. For many systems, there is even less separation. The reason for this lies in the form of equation (3.2). The limiting solubility usually occurs at a low value of the mole fraction. Consequently all realizable values of x are small, and even if n_2^s is large, the product $n_2^s x$ is usually small compared with $n_0 \Delta x/m$. Correspondingly, the value of $(1-x)$ is close to one; hence $n_1^s(1-x)$ is approximately equal to n_1^s, and

$$\frac{n_0 \Delta x}{m} \sim n_1^s. \tag{7.4}$$

The composite isotherm therefore gives an acceptable representation of the individual isotherm for adsorption of the solute, even if there is considerable adsorption of the solvent.

In the past, the distinction between the composite isotherm and the individual isotherm for the solute has rarely been made. It is, however, important to recognize that the distinction must sometimes be made, especially for very soluble solutes. The higher the value of x, the less valid the approximation (7.4) becomes. This is shown in Fig. 7.3, in which the composite isotherm and the individual isotherm for malonic acid diverge markedly as the mole fraction of acid increases. Two effects combine to account for this divergence. One is the high value of x; the other is the relatively high value of n_2^s at high values of x. The importance of the second effect can be seen by comparing Figs. 7.2 and 7.3. For most systems of this kind, the solubility limit occurs at quite low values of x (below 0·1).

THE SIGNIFICANCE OF THE PLATEAU

Relevance of a "Langmuir" Equation

The isotherm in Fig. 7.1 approaches a limiting value of adsorption with increasing concentration. In another type of isotherm (Fig. 7.4), there is a similar region of almost constant adsorption before a final, marked rise.

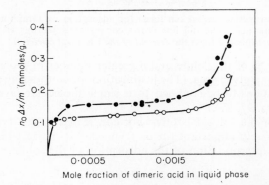

FIG. 7.4. Adsorption of stearic acid from solution in cyclohexane on carbon black[9]: (a) Carbolac I (2400°C), (b) Monarch 74 (2000°C). (Reproduced with permission from the *Journal of the Chemical Society*.)

The resemblance of the curve in Fig. 7.1 to the Type I isotherm for adsorption of a vapour has led many authors to assume, by analogy, that the plateau represents complete coverage of the surface by a monolayer of solute, and that the isotherm should be fitted by a Langmuir equation. These two assumptions need separate consideration.

An equation of the Langmuir type often does fit such isotherms. It will, however, also fit isotherms of the shape shown in Fig. 7.5, which do not form a plateau before the solubility limit is reached.[9] For these four systems, the limits of adsorption given by the equation vary over a four-fold range. They cannot, therefore, all correspond to a complete monolayer coverage of the surface. Thus the applicability of a Langmuir equation is not of itself an

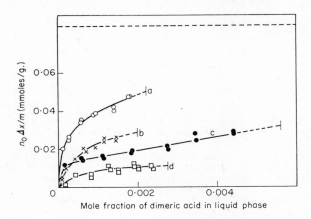

Fig. 7.5. Adsorption of stearic acid from solutions in different solvents on a given carbon black, Spheron 6: (a) cyclohexane, (b) ethyl alcohol, (c) carbon tetrachloride, (d) benzene.[9] The dotted line shows the calculated value for a complete monolayer of stearic acid. (Reproduced with permission from the *Journal of the Chemical Society*.)

indication that monolayer coverage is achieved by any given system. This result is important because, if the isotherm is examined only at low concentrations, it may not be clear whether it effectively reaches a limit (Fig. 7.1) or not (Fig. 7.5).

Reference to Solids of Known Surface Area

The second assumption has to be examined more empirically. For example, Daniel[10] found that on a nickel powder, the adsorption of lauric, palmitic, and stearic acids, and of octadecyl alcohol tended to the same limiting value, expressed in moles/g. This was most readily explained by postulating that the surface was completely covered by a single layer of molecules, each oriented with the major axis perpendicular to the surface. While this approach is valuable, the possibility of chemisorption exists for these particular systems, and thus a more definite result for physical adsorption was required.

This became possible when independent methods were applied to the determination of specific surface areas of suitable adsorbents. The plateau in adsorption of stearic acid on Graphon is the same for four different solvents (Fig. 7.6). It corresponds to an area of 114 sq. Å for each monomeric molecule of adsorbed stearic acid. This is the value which can be calculated

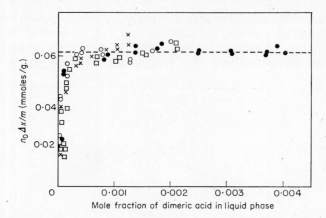

FIG. 7.6. Adsorption of stearic acid on Graphon from: ○, cyclohexane; ●, carbon tetra-chloride; ✗, ethyl alcohol; □, benzene. The broken line shows the calculated value for a complete monolayer of stearic acid.[9] (Reproduced with permission from the *Journal of the Chemical Society*.)

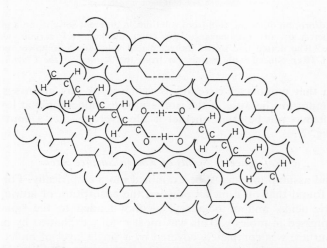

FIG. 7.7. Arrangement of stearic acid molecules in a complete monolayer on Graphon.[9] (Reproduced with permission from the *Journal of the Chemical Society*.)

for the molecules in a complete monolayer packed as shown in Fig. 7.7, with the major axis parallel to the surface.[9] In such a case, therefore, it seems reasonable to suppose that the limit of adsorption corresponds to the completion of a monolayer. Any test less quantitative than this, however, is still open to doubt.

Significance of a Maximum

Although some isotherms show a further rise beyond the plateau, a few are known in which there is a maximum followed by an extensive decline, as in Fig. 7.3. In the case illustrated, the decline can be attributed, in part, to the continuing adsorption of the solvent at high relative concentrations.

The adsorption of the solvent is not the only factor, however. This can be shown by reference to a system in which an effectively complete monolayer of solute is formed at a relatively low concentration which is appreciably below the solubility limit. Subsequently n_1^s is constant and n_2^s is zero. For a relatively high solubility limit, the approximation (7.4) is inadequate, and equation (3.2) must be used. This becomes:

$$\frac{n_0 \Delta x}{m} = n_1^s (1 - x). \qquad (7.5)$$

Thus, after the complete monolayer has been formed, the composite isotherm must fall linearly with concentration. Extrapolation of the linear section to $x = 0$ gives the value of n_1^s corresponding to a complete monolayer.

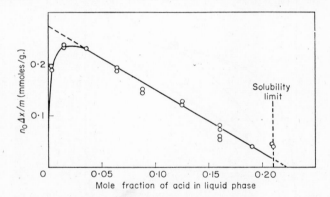

FIG. 7.8. Adsorption on Spheron 6 from aqueous solutions of malonic acid.[11] (Reproduced with permission from the *Transactions of the Faraday Society*.)

Exceptionally, an isotherm is found in which such a linear section occurs, after a maximum which does not correspond to the completion of a monolayer of solute. This can readily be tested because, if a complete monolayer is formed, equation (7.5) requires that the linear part of the isotherm, when extrapolated to $x = 1$, should give a value of zero for $n_0 \Delta x/m$. This clearly does not occur in Fig. 7.8. In such a case it appears that a mixed monolayer of constant composition is formed, i.e. the adsorbed phase contains solvent.

The composition of such a mixed monolayer can be calculated by the method of Schay and Nagy (Chapter 4). In the system illustrated in Fig. 7.8, it seems likely that a "saturated" adsorbed phase is formed which consists of a hydrate of malonic acid instead of anhydrous malonic acid.[11] Other examples of the adsorption of such hydrates are referred to below (p. 122).

Absence of a Limiting Value

Although isotherms tending to an adsorption limit are very common, other forms of isotherm have been observed. In some cases, a very definite plateau is followed by a further rise close to the solubility limit. This is difficult to measure accurately, but appears to be definite in many cases. It probably corresponds to incipient crystallization of the solute.[9]

Figure 7.9 shows the adsorption of benzoic acid from water by several carbons.[12] Limiting adsorption is found for adsorption by carbons with narrow pores (a, d, and e). For the highly activated Saran carbon (b), which

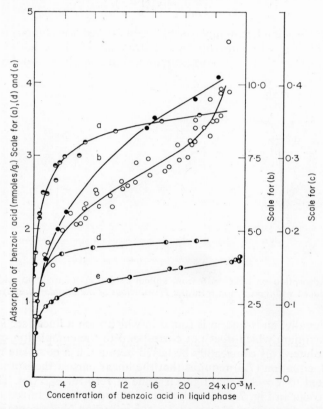

FIG. 7.9. Adsorption by several carbons from aqueous solutions by benzoic acid: (a) steam-activated nutshell charcoal, (b) steam-activated Saran charcoal, (c) Graphon, (d) unactivated Saran charcoal, (e) decolorizing charcoal (Darco G 60). (Reproduced with the permission of the American Chemical Society from the *Journal of Physical Chemistry*.)

probably has a large volume present as wide pores, the isotherm does not appear to tend to a limit. For adsorption on Graphon (c), which is non-porous, a sigmoid isotherm is found, suggestive of multilayer formation. The rather ill-defined "knee" would correspond to a complete monolayer of benzoic acid adsorbed with the plane of the benzene ring and of the carboxyl group parallel to the surface. The subsequent rise in the isotherm would correspond to multilayer formation, the recorded data giving no indication of a limit to this process.

THE EFFECT OF TEMPERATURE

With rise in temperature, the isotherm usually falls to lower levels, particularly at the lowest concentrations, though it may reach almost the same limiting value at high concentrations (Fig. 7.10). This corresponds to a

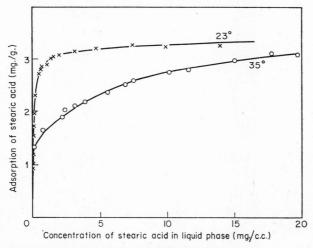

FIG. 7.10. Effect of temperature on adsorption of stearic acid from benzene by nickel powder.[10] (Reproduced with permission from the *Transactions of the Faraday Society*.)

weakening of the attractive forces between the solute and the solid surface (and between adjacent adsorbed solute molecules) with increasing temperature, and corresponding increase in solubility of the solute in the solvent. Thus if the solute is regarded as distributed between the adsorbed layer and the solution in a partition equilibrium, the position of equilibrium is displaced in favour of the solution as the temperature rises.

For such systems it has been suggested (Chapter 5) that the important parameter is the reduced concentration (relative concentration) as being a measure of activity or escaping tendency of the solute. In the simplest cases, it is found that isotherms for a given system but for different temperatures can be superimposed if plotted against relative concentration, e.g. in adsorption of iodine from cyclohexane by Graphon (Fig. 7.11). Such close agreement is not found if other solvents are used, nor with

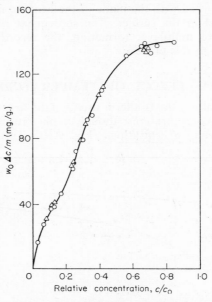

FIG. 7.11. Adsorption of iodine from cyclohexane by Graphon, plotted as a function of relative concentration: $\bigcirc$, 20°c; $\triangle$, 40°c.

Spheron 6 as the adsorbent instead of Graphon. Effects other than the activity of the solute are thus important, though this may well be the dominant effect.[13]

CHEMISORPTION

Chemisorption has been found to occur in a number of systems during adsorption at room temperature. The most extensive investigations have been made with fatty acids and oxides, including the oxide films frequently found on the surfaces of metals, but several quite different systems have also been investigated.

As in chemisorption of gases, it is rarely possible to show convincingly that chemisorption from solution occurs, by reference to one criterion only. As many criteria as possible should be considered.

The Shape of the Adsorption Isotherm

A particularly clear case is shown in Fig. 7.12, in which there is no indication that the isotherm rises from the origin.[14] It appears to start from the adsorp-

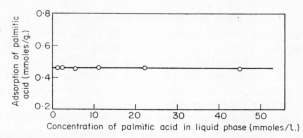

FIG. 7.12. Chemisorption of palmitic acid (from solution in benzene) by Adams platinum catalyst.[14] (Reproduced with permission from the *Journal of the American Chemical Society*.)

tion axis at a value which corresponds to the degree of chemisorption. The amount of solute adsorbed would be expected to be independent of concentration, except at the lowest concentrations, at which the amount of solute in the volume of solution used would be less than the capacity of the surface. The constant level of the isotherm accords with this; it also suggests that chemisorption is not accompanied by preferential physical adsorption in this case. The level of adsorption is also independent of the solvent used, and of the chain-length of solute in a homologous series.

More generally, preferential physical adsorption is likely to occur in a second layer on top of the chemisorbed material in the first layer. The observed isotherm is then the sum of two isotherms, one dependent and one independent of the concentration of solution. The total isotherm then rises with concentration, but appears to originate at a point on the adsorption axis well above the origin,[15] as in Fig. 7.13.

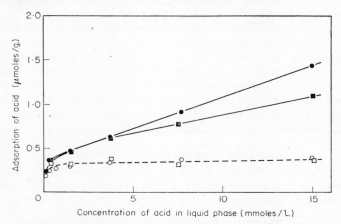

FIG. 7.13. Chemisorption and physical adsorption of capric and stearic acids from benzene on steel: ○, stearic acid; □, capric acid.[15] (Reproduced with the permission of the American Chemical Society from the *Journal of Physical Chemistry*.)

Time Required for Adsorption

At room temperature, chemisorption is usually a much slower process than physical adsorption. Consequently a much longer time may elapse before a steady concentration is reached in the solution. A very marked example of this is found in the adsorption of stearic acid from benzene on metal powders.[10] The metals can be divided into two groups according to the time taken for adsorption "equilibrium" to be reached, i.e. for a steady concentration in the solution to be reached. The difference between the curves in Fig. 7.14(a) and those in Fig. 7.14(b) suggests the difference between physical adsorption only and physical adsorption accompanied by chemisorption, though this would have to be confirmed by other methods.

The time involved in such experiments depends not only on the nature of the solute (mainly on its mobility in solution) but also on the porosity of the adsorbent. In a given case, therefore, it is useful to ascertain how long is required for equilibrium to be established when the adsorbent is brought into contact with a solution from which physical adsorption only can occur.

Desorption

The rate of desorption with an appropriate solvent is similarly a useful guide. More important is the observation as to whether some solute remains on the surface of the adsorbent after the rate of desorption has fallen to zero (strictly, to a value too low to be detected). In this respect, adsorbates of very high molecular weight may constitute a special case, as is discussed in the next chapter. More definite conclusions can be drawn if the desorbed material is not the original adsorbate but a derivative formed by reaction between the adsorbate and the adsorbent. Thus a soap is frequently desorbed after a fatty acid has been chemisorbed by a metallic oxide.

It may be necessary to distinguish between chemisorption (which ceases to deplete the solution once the surface is saturated) and slow chemical reaction involving the whole of the solid. Thus a blue coloration has been observed in a solution of stearic acid in benzene when the solution has been brought into contact with copper powder.[10] This must be due to desorption of a copper soap, implying at least reaction of the stearic acid with a substantial film of oxide on the surface of the metal, and possibly with the metal itself.

Molecular Area of Adsorbed Material

In the simplest cases, the area occupied by physically adsorbed molecules is determined by the packing which is possible in a single layer of molecules, i.e. the spacing is determined by the size of the molecules themselves. In chemisorption, the spacing is determined by the availability of the sites at which chemical bonding can take place. On amorphous materials this is likely to be irregular, and leads to a higher molecular area than would be expected for close-packing. For crystalline materials, it may be possible to

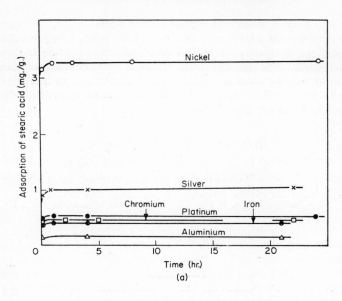

(a)

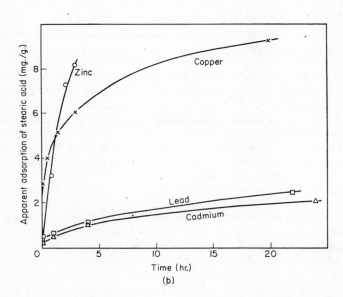

(b)

FIG. 7.14. Adsorption of stearic acid from benzene by metal powders as a function of time.[10] (Reproduced with permission from the *Transactions of the Faraday Society*.)

calculate the effective molecular area, which is normally greater than corresponds to close-packing. This may be due to wide spacing of the active sites. Alternatively the sites may be spaced so closely that a molecule adsorbed on one site partly overlaps and thus blocks a neighbouring site; the next molecule is therefore adsorbed on the third site, which may be further from the first site than corresponds to the closest distance of approach of the molecules. Examples are given on p. 114.

Detection by Special Techniques

It may be expected that physical methods of examining molecular structure (e.g. infra-red spectroscopy) will increasingly be used for examining substances adsorbed from solution.

An interesting special case is the use of electron spin resonance for determining the state of polycyclic hydrocarbons adsorbed from solution in simple organic solvents onto silica-alumina cracking catalysts. Perylene was shown to be adsorbed as a positive ion.[16] The adsorption of both perylene and anthracene was accompanied by transfer of one electron from each molecule to a Lewis acid site on the surface.[17] A weak spectrum was shown by naphthalene, but benzene and its methylated derivatives showed no effect. Desorption was readily effected by addition of water.

In a quite different investigation it was suggested that phenanthrene is adsorbed from solution onto alumina by forming a π-bonded complex with aluminium ions.[18] This is referred to further below (p. 120).

ORIENTATION OF THE ADSORBED MOLECULES

The orientation of the adsorbate generally presents no problem in chemisorption (at least for mono-functional adsorbates), as the functional group determines the point of attachment. Thus the long-chain fatty acids are attached to the surface by the carboxyl group, with the hydrocarbon chain perpendicular to the surface. Consequently, the same number of moles are chemisorbed by the same weight of a given adsorbent for different members of a homologous series, though the extents of physical adsorption may differ;[15] this is shown in Fig. 7.13.

It has sometimes been assumed that this orientation is normally adopted at the liquid–solid interface, even in physical adsorption. The assumption reflects the dominance exerted by the elegant experiments made with compressed monolayers on aqueous substrates and the comparative paucity of results for adsorption on solids. It is now beginning to appear that in physical adsorption, a different orientation is quite common. In adsorption on carbon blacks, the results described above (p. 97) suggest that stearic acid is adsorbed with the hydrocarbon chain parallel to the surface. This is also true of other acids in the homologous series,[19] and may be true generally of the adsorption of these acids by non-polar solids.[20] It should also be recalled that, even on aqueous substrates, these fatty acids are thought to adopt an orientation parallel to the surface at low compression.

The parallel orientation on non-polar solids is readily understood, because the fatty acids are largely dimerised in such solvents as benzene. In the parallel orientation the double hydrogen-bonding between the carboxyl groups of the dimer is preserved on adsorption. In the perpendicular orientation, however, these bonds must be broken, with adsorption of the monomer. When the adsorbent is not polar, no strong bond can be formed to compensate for the rupture of the hydrogen-bonds in the dimeric molecule. For the highly polar aqueous substrate such compensation is evidently available, and it is therefore interesting to discover how polar a solid must be for the acid to be adsorbed as monomeric molecules perpendicular to the surface.

In early work on oxides, it was suggested that oleic and butyric acids adopted the perpendicular orientation on titania,[21] as did stearic acid on aluminium hydroxide (alumina trihydrate).[22] In these experiments it was not clear whether adsorption was physical or chemical in character. This now seems an important distinction to draw, especially with basic solids.

Results for adsorption of stearic acid on a range of non-porous solids of known surface area have shown that reversible adsorption can occur with the perpendicular orientation on alumina and titania.[20] Zettlemoyer has shown that titania has a much higher surface polarity than silica,[23] and the same may be expected to be true of alumina. The range of polarity which can be investigated is limited. More highly polar solids (e.g. magnesia) undergo bulk reaction with the fatty acids instead of adsorption. The special situation of metals as adsorbents is considered below.

Linear dibasic acids are able readily to adopt the parallel orientation, because hydrogen-bonding through the carboxyl groups forms chains of infinite length (see p. 122). In such cases the adsorbed layer may be compared with a single layer of the corresponding crystal. It is even possible for the adsorbed layer to incorporate water molecules regularly in the structure, rather as they are incorporated in a crystal to form a hydrate.

Paraffin molecules, with no functional groups, can only be expected to adopt a parallel orientation to any surface. This has been found experimentally in adsorption from carbon tetrachloride on carbon blacks.[24]

Other substances have molecules intermediate in polarity between paraffins and fatty acids. Daniel's results[10] for the adsorption of octadecyl alcohol from benzene on metal powders show a sharp break in the isotherm (Fig. 7.15). A similar effect is shown by ethyl stearate. In each case, the two levels of adsorption are in the ratio which corresponds to the areas occupied by the molecules in the two extreme orientations. It seems, therefore, that the molecules are adsorbed parallel to the surface from solutions of low concentration, but perpendicular to the surface from solutions of high concentration. On alumina, however, long-chain alcohols appear to be adsorbed perpen-

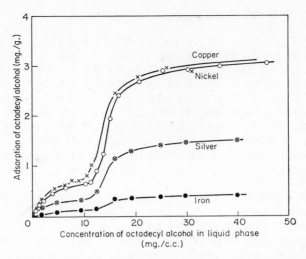

FIG. 7.15. Adsorption of octadecyl alcohol from benzene by metal powders[10] at 23°C. (Reproduced with permission from the *Transactions of the Faraday Society*.)

dicular to the surface.[25] Although the authors did not consider the possibility, McBain and Dunn's results for adsorption of cetyl alcohol by magnesium oxide[26] are probably best interpreted in terms of orientation parallel to the surface.

On silica, however, octanol appears to be adsorbed with a parallel orientation, but shorter alcohol molecules probably tend towards the perpendicular orientation. Thus on three samples having different degrees of hydration the limiting adsorption per unit area of surface was as follows:

Sample	A	B	C
Water content (%)	7·9	4·2	2·8
Limiting adsorption (micromoles/sq. m.):			
CH_3OH	13·5	9·5	8·5
C_3H_7OH	5·0	4·8	4·4
$C_6H_{13}OH$	3·6	3·4	3·2
$C_8H_{17}OH$	2·6	2·6	2·6

The variation in adsorption for the lower alcohols is related to the water content and hence to the surface concentrations of hydroxyl groups. Adsorption is then specific to these sites.[27]

"STEPPED" ISOTHERMS

In general, composite isotherms show a smooth variation with composition of solution. In some cases, however, "steps" or sharp discontinuities have been observed. These arise for more than one reason. In Chapter 4, a single discontinuity in the isotherm in adsorption from a completely miscible pair of liquids was attributed to the appearance of a second layer of adsorbate on top of the first. A similar phenomenon has been observed in the adsorption of decylamine from water by platinum,[28] though in this case, a smooth step rather than a sharp break is observed (Fig. 7.16). The first layer is

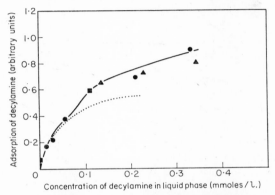

FIG. 7.16. Adsorption of decylamine on platinum from aqueous solution[28] at pH 10·6–10·7. (Reproduced with the permission of the American Chemical Society from the *Journal of Physical Chemistry*.)

probably adsorbed with the polar group towards the platinum and the second layer with the polar group towards the solution. A single break, in the adsorption of lauric acid from carbon tetrachloride by carbon black, appears to be due to a phase change in the adsorbed acid.[19] A sharp break in the isotherm can also be due to a change in the orientation of the adsorbate if this is adsorbed strongly enough to form a complete monolayer at the lower concentrations (Fig. 7.15).

In some cases two and even three sharp discontinuities are observed. Venturello has published a series of papers in which he presents such isotherms for adsorption by inorganic solids—alumina, silica gel, magnesium hydroxide, and calcium carbonate—from solutions of dyestuffs in water, and of iodine in organic solvents.[29–32] A well-developed example is shown in Fig. 7.17 in which the contrast between continuous and discontinuous isotherms is apparent. With a given adsorbent and solute, discontinuities are found with some solvents but not with others. In these cases the discontinuous nature of the isotherm seems to be well established. The explanations, however, are not altogether clear. It is suggested that

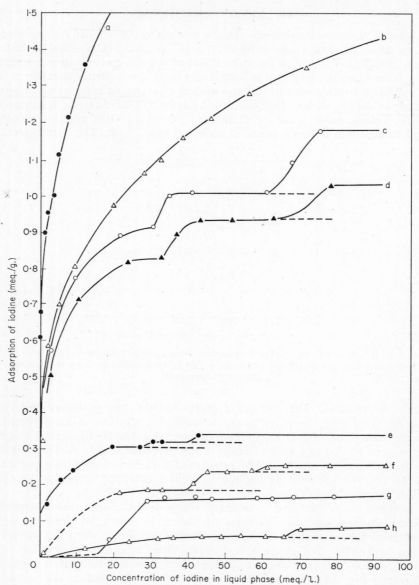

FIG. 7.17. Adsorption of iodine from solution onto magnesia prepared at 400°C (upper curves): (a) in pentane, (b) in carbon tetrachloride, (c) in dioxan, (d) in benzene; adsorption of iodine from solution onto magnesia prepared at 700°C (lower curves): (e) in pentane, (f) in carbon tetrachloride, (g) in dioxan, (h) in benzene.[31] (Reproduced from *Gazzetta*.)

the surfaces of the adsorbents are heterogeneous and that different parts of the isotherm correspond to adsorption on chemically different sites (sites of differing energy). It remains, however, to identify the chemical nature of the different sites (it is more difficult to visualize three than two such sites for the inorganic solids in question). The effect of the solvent is also important. In the adsorption of iodine by magnesia, discontinuous isotherms were observed in adsorption from pentane, carbon tetrachloride, dioxan, and benzene if the magnesia was prepared at 700°C, but only in adsorption from the last two solvents if it was prepared at 400°C. Venturello suggests that this may be related to the formation of molecular association compounds between iodine and both dioxan and benzene. Moreover, as very few solid surfaces are really homogeneous, it would be interesting to know why discontinuous isotherms are not observed more frequently. Possibly the different parts of the surface normally differ only slightly in adsorbing power for most systems.

Other systems which give rise to stepped isotherms are discussed below (pp. 116, 118).

BEHAVIOUR OF INDIVIDUAL SYSTEMS

Monocarboxylic Acids

The importance of long-chain fatty acids in the study of lubrication of metal bearings has led to numerous experiments on the adsorption of these materials and of related compounds (e.g. alcohols, esters, and amines) not only on metal surfaces, but also on other solids. Hydrocarbons have usually been used as solvents.

(a) Adsorption on Metals

Two features are particularly important in adsorption on metals. First, many metals quickly acquire a thin layer of oxide (or hydroxide) even on freshly-prepared surfaces. A distinction must therefore be made between adsorption on "clean" metal surfaces and on oxide coatings. Much early work must have been carried out with the latter. Secondly, chemisorption is very common on metal surfaces of either kind, and often outweighs physical adsorption in importance. In some cases, chemical reaction goes beyond the formation of a chemisorbed monolayer, with production of metal soaps which are found in the liquid phase. "Clean" metal surfaces have been prepared by machining the metal under an inert liquid such as cyclohexane and carrying out adsorption experiments without allowing the metal subsequently to come into contact with air.[33]

The more electropositive metals (e.g. magnesium, aluminium, nickel, lead, copper, cobalt, and indium) when prepared with clean surfaces in this way, adsorb n-nonadecanoic acid from cyclohexane, apparently giving a complete monolayer.[34, 35] If it is assumed that there is no surface roughness, the limiting adsorption corresponds to a molecular area of 24 sq. Å in each case. The acid can be removed as a soap, and is thus presumed to be chemisorbed. The adsorbed layer is virtually close-packed, with the major axis of the hydrocarbon chain perpendicular to the surface. The adsorbed film is not always static.[35] If the metal soap is desorbed, further attack by the acid on the freshly exposed metal may take place. If desorption of the soap is fast relative to the rate of chemisorption, as with aluminium, the level of adsorption may appear to be low.

A lower degree of adsorption is observed with the less electropositive metals (silver, gold, and platinum), and the acid has been recovered as such from gold surfaces. These observations are compatible with physical adsorption.

As "clean" surfaces are not normally encountered with the more reactive metals, it is useful to examine the adsorptive behaviour of the metal as normally prepared. The oxide film may be either thick and permeable or compact.[36] Thick films appear to "adsorb" fatty acids steadily for a period of several days (in contrast to "clean" surfaces at which equilibrium is reached rapidly). This process is better regarded as a continuing chemical reaction than as adsorption.[33] Soap formation may be sufficiently extensive with copper to colour the solution blue.[10] In other cases, the occurrence of reaction is demonstrated by using radioactive samples of metal and subsequently detecting radioactivity in the solution. In this way Bowden and Moore showed that gold and platinum adsorb long-chain compounds without reaction.[37] Zinc, cadmium, and copper react strongly with stearic acid.

In some cases (e.g. Raney nickel and Adams platinum catalyst) chemical reaction is confined to the formation of a monolayer of fatty acid.[14] The adsorption isotherm is then a straight line (Fig. 7.12) with no evidence of a fall towards the origin. If a sample of metal is placed in a solution containing less acid than is required to form a monolayer, the available acid is completely removed from solution. Adsorption is independent of the solvent (provided that the solvent does not react with the metal), and of the chain-length of the acid over the range C_{10} to C_{22}.

Oxide surfaces on metals differ considerably in their behaviour, and different degrees of reactivity can be demonstrated not only with different metals, but also with different oxide films on the same metal.[38] In some cases, particularly for the more reactive powders, the behaviour towards the fatty acids can be altered considerably by small traces of water.[39] Physical adsorption sometimes occurs on chemisorbed films (Fig. 7.13). Hackerman has shown the possibility of distinguishing between "total adsorption" and "firm adsorption", the latter referring to acid which is not removed by washing the metal with benzene.[40]

The extent of chemisorption of lauric, palmitic, and stearic acids on steel is reported to be insufficient to form a close-packed monolayer.[15] This, however, was presumably judged by standards appropriate to physical adsorption, whereas a lower level is expected for adsorption on specific sites (see p. 104). Caproic and benzoic acids reacted continuously with the steel.

On some metals, used as powders without removal of any oxide film potentially present, reversible adsorption of fatty acids occurs.[10] The isotherms are then as shown in Fig. 7.18. They reach a limiting value and can be fitted by a Langmuir equation (but not by a Freundlich equation), if this is expressed in terms of solute concentration.

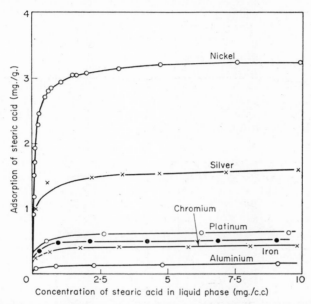

FIG. 7.18. Reversible adsorption of stearic acid from benzene by metal powders.[10] (Reproduced with permission from the *Transactions of the Faraday Society*.)

The significance of a limit, as Daniel pointed out,[10] is not clear. Before it can be elucidated, two prior issues must be settled: first, whether the adsorption of the solvent can be ignored; second, why adsorption reaches a limiting value. As the limiting adsorption of octadecyl alcohol on these solids (Fig. 7.15) is approximately the same as that of stearic acid, it seems likely that a complete monolayer is formed in each case, with the hydrocarbon chain perpendicular to the surface.

(b) Adsorption on Oxides and Hydroxides

This group of adsorbents includes oxides, hydroxides, and compounds of intermediate composition (e.g. alumina monohydrate, $AlO.OH$). In some cases authors have not specified the precise composition of the solid.

In adsorption of fatty acids by silica, Claesson found a variation according to the solvent used, polar solvents competing for the surface, but non-polar solvents showing little competition.[41] When non-polar solvents were used, the same isotherm was found for a range of fatty acids when the results were expressed in terms of weight adsorbed by unit weight of solid. This implies that the acids were adsorbed with the major axis parallel to the surface.

On various aluminas, however, lauric acid was found to be adsorbed with the major axis perpendicular to the surface. The molecules were not close-packed, and this implied adsorption on specific sites at the surface (oxygen atoms or ions), the spacing of the adsorbed acid molecules being determined by the spacing of the lattice.[42] These results have been confirmed in experiments with non-porous solids of known specific surface area and have been extended to titania, on which stearic acid adopts the perpendicular orientation.[20] The orientation thus depends on the polarity of the surface. On the relatively non-polar carbon and silica surfaces, the parallel orientation is adopted, with close-packing. On the more polar alumina and titania, the perpendicular orientation is adopted, but with spacing related to that of the oxide or hydroxide groups in the surface.

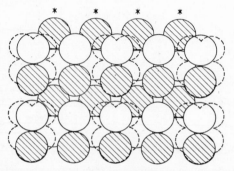

FIG. 7.19. Adsorption of stearic acid (dotted outline) on the *a* plane of γ-alumina monohydrate. The shaded circles represent hydroxide ions, and the unshaded circles represent oxide ions. Rows marked * are at a lower level than their neighbours.[20] (Reproduced with permission from the *Journal of the Chemical Society*.)

On the non-polar solids the fatty acids are probably adsorbed as dimeric molecules, this being the predominant form in solutions in non-polar solvents. There is no group on the surface to which the carboxyl group of the fatty acid could be attached in preference to remaining hydrogen-bonded to another carboxyl group to form a dimer. On polar surfaces, however, the most effective use is made of the surface by the hydrogen-bonding of a large number of monomeric molecules to the surface instead of van der Waals adsorption of a much smaller number of dimeric molecules. In this process hydroxide ions are more effective than oxide ions.[43]

The way in which adsorption on sites prevents close-packing is shown in Fig. 7.19. For the surface shown, the spacing of the hydroxide ions is too great to allow close-packing along the rows. The spacing between the rows is too small for adsorption to occur on the oxide ions. Thus only a relatively small fraction of the surface appears to be covered. In other cases, however, adsorption on sites differs very little from close-packing (Fig. 7.20). When the nature of the crystal lattice is known, the "coverage" can be calculated approximately, and is found to agree well with the experimental values.[20] Adsorption of stearic acid on water[44] can be regarded as a special case of adsorption by hydrogen-bonding to specific sites, in which the surface of the substrate is mobile, the water molecules being able to take up positions which allow close-packing of the fatty acid molecules when the film is compressed.

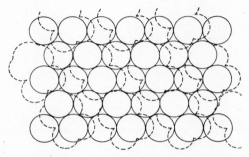

FIG. 7.20. Adsorption of stearic acid (dotted outline) on close-packed oxide or hydroxide ions.[20] (Reproduced with permission from the *Journal of the Chemical Society*.)

It would be interesting to examine adsorption on more polar oxides than those already mentioned. No great extension of the range of polarity is possible, however, because bulk reaction instead of adsorption occurs, e.g. with magnesia. Even with alumina, chemisorption of stearic acid occurs slowly at room temperature, and aluminium stearate can be desorbed. In the initial stages the stearic acid is thought to be adsorbed by hydrogen-bonding. Adsorption on silica and on titania appears to be completely reversible at room temperature.[20]

(c) *Adsorption on Carbons*

The adsorption of stearic acid on non-porous carbons of known specific surface area has been referred to earlier in this chapter (pp. 96–98). When the surface is free of oxygen complexes, a close-packed monolayer is readily formed, with the molecules, probably present in dimeric form, adsorbed with the major axis parallel to the surface. On surfaces bearing oxygen complexes, competition from the solvent prevents the formation of a complete monolayer of acid even at the highest available concentrations.

Similar generalizations can be made about the adsorption of the lower members of the homologous series which are solid at room temperature.[45] For a given adsorbent and solvent the lower acids give a lower surface coverage than the higher acids at a given concentration. Some isotherms are specific to an individual system. For example, two stages are observed

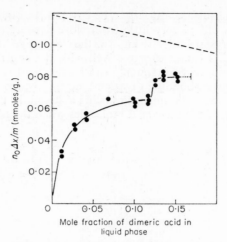

FIG. 7.21. Adsorption of lauric acid from carbon tetrachloride on Graphon. The broken curve shows the value of $n_0 \Delta x/m$ for a complete monolayer of lauric acid.[19] (Reproduced with permission from the *Journal of the Chemical Society*.)

in the adsorption of lauric acid from carbon tetrachloride on Graphon (Fig. 7.21). The break in the isotherm would be expected to correspond to a phase change in the adsorbed layer, but the nature of this change is not yet clear.

Oleophobic Films

In some cases the adsorption of long-chain compounds on solid surfaces leads to the formation of an oleophobic film which is not wetted at room temperature either by the original solution or by the pure solvent. Examples are the adsorption of n-eicosyl alcohol, primary n-octadecylamine, and n-nonadecanoic acid from hexadecane on platinum and on Pyrex glass. Such films have been described as "autophobic".[46]

The phenomenon has been studied extensively by Zisman.[47] It is attributed to the orientation of the adsorbed molecules, at high surface coverage, with the hydrocarbon chains directed away from the solid surface. In effect, a new surface of methyl groups is presented to the solution, and if this is sufficiently regular, it is hydrophobic. There is evidence that oleophobic films are most readily formed on exceptionally smooth surfaces.[48] The films can be wetted at higher temperatures, and the temperature of wetting varies according to a number of factors, including chain-length, polar group, and solvent.

Alcohols, Esters, and Amines

Irreversible adsorption on steel has been observed for alcohols, esters, and amines,[15] as well as for fatty acids. The number of molecules of a given type irreversibly adsorbed by unit area of surface is independent of molecular weight. As between the chemical species, the amount decreases in the order acid > amine > alcohol > ester. The quantity physically adsorbed onto the layer adsorbed irreversibly varies according to the molecular weight of the compound.

Bowden and Moore showed that zinc, cadmium, and copper do not react with octadecyl alcohol.[37] An apparent slight reaction with ethyl stearate was probably due to the presence of acid either as an initial impurity or as a product of hydrolysis by traces of water unintentionally present during the experiment.

In physical adsorption on metals, the alcohols and esters are less strongly adsorbed than the corresponding acids. The isotherms for adsorption of

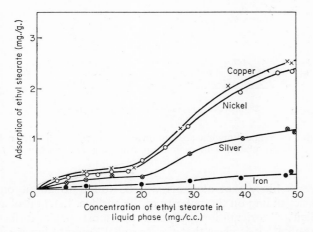

FIG. 7.22. Adsorption of ethyl stearate from benzene on metal powders[10] at 23°c. (Reproduced with permission from the *Transactions of the Faraday Society*.)

octadecyl alcohol from benzene on copper, nickel, silver, and iron are shown in Fig. 7.15. The characteristic feature is the existence of two steps, which are explained in terms of different orientations of the alcohol molecule when adsorbed from solutions of low and high relative concentration, respectively (cf. p. 107). Similar behaviour is shown by ethyl stearate when adsorbed from benzene on the same metal powders (Fig. 7.22). In this case the step in the isotherm is less sharply defined, and the isotherms do not appear to reach a limiting value at the highest concentration studied. This may simply be an effect of temperature, as the same shape of isotherm is shown by octadecyl alcohol at a higher temperature (Fig. 7.23). As the ester has a lower melting point (30·6°c) than the alcohol (58·0°c), a lower temperature might well be required for the two steps in the isotherm to be defined sharply.

Competition for the surface from the solvent has been found in the adsorption of cetyl alcohol on magnesium oxide, the extent of the effect differing according to the solvent.[26]

The orientation of alcohols and esters in the adsorbed phase has been examined less extensively than that of acids. From Daniel's evidence for

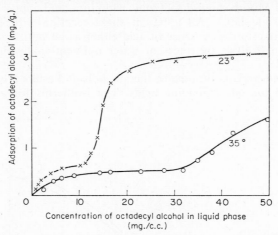

FIG. 7.23. Effect of temperature on adsorption of octadecyl alcohol from benzene on nickel powder.[10] (Reproduced with permission from the *Transactions of the Faraday Society*.)

metals it seems likely that alcohols will be found to adopt one of two orientations depending on the polarity of the surface. Esters probably adopt a "perpendicular" orientation only on the most polar surfaces. From Claesson's results[41] it is clear that the "parallel" orientation is adopted on silica.

Hydrocarbons

(a) Alkanes

In the light of the above discussion, the adsorption of hydrocarbons is interesting because no polar group is present in the molecules. The normal alkanes are further interesting because they are soluble in simple organic solvents to a useful extent at higher chain-length than the corresponding polar compounds (fatty acids or alcohols). The higher members partly bridge the gap between simple organic molecules and polymeric materials which show different adsorptive behaviour, discussed in Chapter 8.

The isotherm for adsorption of the long-chain alkanes from carbon tetrachloride on Graphon shows a very well defined plateau (Fig. 7.24). The level of adsorption at the plateau is approximately the same for $C_{18}H_{38}$, $C_{22}H_{46}$, $C_{28}H_{58}$, $C_{32}H_{66}$, and $C_{36}H_{74}$, when measured in units of mg./g. This implies that the molecules, in each case, are oriented with the major axis parallel to the surface. The level of adsorption shows that the molecules are also close-packed.[24] The pattern of adsorption is thus very similar to that of the long-chain fatty acids.

The preferential adsorption of the alkanes on Spheron 6 was found to be too small to be measured accurately, as is shown by the scatter of points

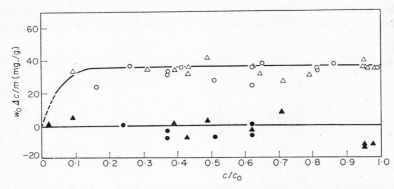

FIG. 7.24. Adsorption of n-octadecane and of n-docosane from carbon tetrachloride at 20°c on Graphon and on Spheron 6: ○, n-octacosane on Graphon; ●, n-octacosane on Spheron 6; △, n-docosane on Graphon; ▲, n-docosane on Spheron 6.

in Fig. 7.24. This again confirms the pattern of behaviour of the fatty acids, namely that the solvent competes effectively for the more polar sites on the solid surface. Competition with the alkanes is more effective than with the fatty acids, as the former have no polar groups.

A further factor which may be important is that the irregular nature of the surface of Spheron 6 would tend to disturb regular close-packing in a quasi-crystalline layer, which is thought to occur on Graphon. Lateral interactions between the molecules would tend to stabilize the layer adsorbed on Graphon, but could presumably occur only to a small extent in the layer adsorbed on Spheron 6. Further evidence for this view is that preferential adsorption of n-tetradecane from carbon tetrachloride on Graphon is negligible, and that of squalene and squalane from cyclohexane is small. Although these substances are of considerable molecular weight, they are liquids and could not form a quasi-crystalline adsorbed layer.

The desorption of the higher n-alkanes from Graphon by cyclohexane and by carbon tetrachloride occurs very slowly indeed at room temperature. Complete desorption can be achieved, however, with hot solvent in a period of 12 hours. The alkane molecules are thus firmly held by Graphon. On the other hand, they are not firmly held by Spheron 6, as the solvent competes very effectively for the surface. This suggests that each methylene group is held relatively weakly by the surface. The difficulty of desorption is thus due to the additive effect of adsorption forces operating at a large number of groups simultaneously—considerably greater than the number in cyclohexane, used as a desorbing agent.

(b) Aromatic Hydrocarbons

A less common shape of isotherm is found in the adsorption of aromatic hydrocarbons from xylene on chromatographic alumina.[18] The curve is convex to the concentration axis at low concentrations (Fig. 7.25), and may or may not reach a limiting value. The limiting value for phenanthrene

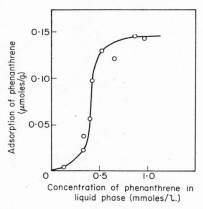

FIG. 7.25. Adsorption of phenanthrene from solution in xylene on alumina[18] at 20°C. (Reproduced with permission from the *Journal of the Chemical Society*.)

corresponds to coverage of a very small fraction of the surface (0·06%). From the shape of the isotherm (cf. p. 129), Giles argues that the molecules are probably adsorbed with the plane of the rings perpendicular to the surface, small clusters being formed on active sites. These latter may be aluminium atoms exposed on the outer surface of the adsorbent as a result of mechanical damage.

The effect of substituents deserves further study. Benzene is strongly adsorbed from solution in cyclohexane by charcoal. At corresponding concentrations (moles per litre), the adsorption of substituted benzenes (mesitylene, *o*- and *p*-xylene, t-butyl benzene, isopropyl benzene) is very much less.[49]

Dicarboxylic Acids

The adsorption of dicarboxylic acids is interesting in the light of a significant feature found in the adsorption of monocarboxylic acids; the acid molecules tended to remain in the dimeric form unless the adsorbent had a very polar surface. The dicarboxylic acids can be studied in aqueous solution in which they may well be monomeric but hydrated. The adsorbed phase, however, might be analogous with the crystal, in which infinite chains can be formed.

Some interesting experiments were carried out by Morrison and Miller.[50] For a series of porous charcoals, the maximum adsorption from aqueous solution was found to alternate for successive members of the homologous series (Fig. 7.26). This alternation closely parallels the alternation in the melting points of the acids. It was attributed to rotation of the adsorbed molecules, the acids with an even number of carbon atoms (and a *trans* configuration of the carboxyl groups) occupying a smaller area than those with an odd number (and a *cis* configuration).

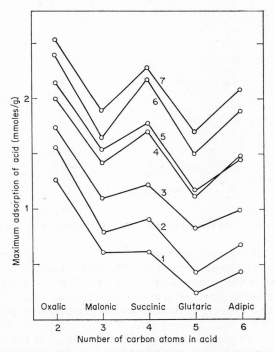

FIG. 7.26. Maximum adsorption of dicarboxylic acids on an activated series of charcoal (nos. 1–7).[50] (Reproduced with permission from the *Canadian Journal of Chemistry*.)

A difficulty in assessing this work is that there may have been molecular sieve action in the charcoals studied, which seems likely as the extent of alternation changed on progressive activation of the charcoals. This would affect the penetration of the "odd" acids (with maximum width of about 5·9 Å) more than the "even" acids (with maximum width of about 5·15 Å).

The adsorption of these acids on non-porous carbon blacks has been studied in more detail.[45] Three shapes of isotherm were observed. Those with a maximum (see Fig. 7.3) have been discussed above, and occur only with malonic and glutaric acids. The other moderately soluble acids gave an isotherm similar to that shown in Fig. 7.1. Some of the sparingly soluble acids gave isotherms with a remarkably long rising section at high relative concentrations (Fig. 7.27).

For the series as a whole, the level of adsorption at the maximum or at the plateau was too small to be interpreted in terms of adsorption with

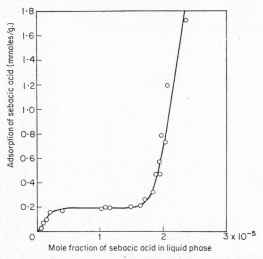

FIG. 7.27. Adsorption on Graphon from aqueous solutions of sebacic acid at 20°C.

the major axes of the molecules perpendicular to the surface. The results were therefore considered in terms of "parallel" orientation, which gives slightly different arrangements for "odd" and "even" acids (Fig. 7.28a and b). The former are packed less tightly than the latter, so that a given "odd" acid has an effective molecular area in the adsorbed phase almost as great as that of the next "even" acid. The molecules are assumed to be hydrogen-bonded to give chains of infinite length as in the corresponding crystals. Even with these orientations, the formation of a complete mono-layer was not observed for the lower members (Fig. 7.29). An alternation is shown, but in the opposite sense from that found by Morrison and Miller. In this case it is thought to be related to the alternation in solubilities of the acids, the "odd" acids being the more soluble.

As Graphon is hydrophobic, it seems unlikely that normal competition from water for the surface is the cause of failure of the lower acids to form complete monolayers. It therefore seems likely that the water molecules present in the adsorbed layer must be bonded to the acid molecules. A net-work of hydrogen-bonds would retain the water molecules in the adsorbed layer. For oxalic acid, the adsorbed layer would be similar to a single layer of the crystalline dihydrate, though probably incorporating more water.

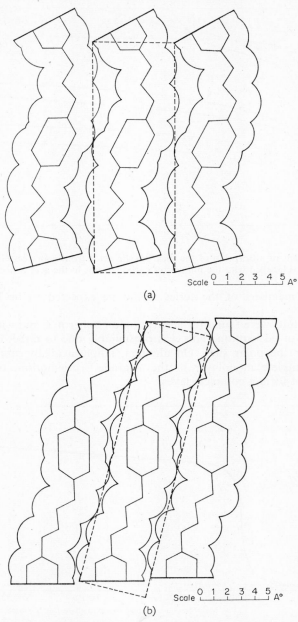

Scale 0 1 2 3 4 5 A°

(a)

Scale 0 1 2 3 4 5 A°

(b)

FIG. 7.28. Arrangement of (a) glutaric acid (C_5) molecules, and (b) adipic acid (C_6) molecules in adsorbed layers. The rectangles indicate the effective area of two molecules of each acid.

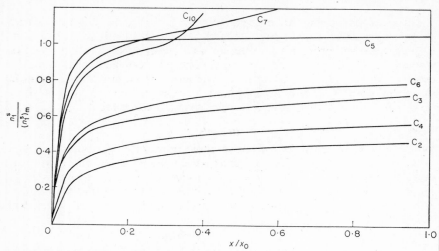

Fig. 7.29. Extent of adsorption of the dicarboxylic acids from water on Graphon. Each curve is labelled with the number of carbon atoms in the acid concerned.

The higher members of the series would be expected to be less heavily hydrated.

Similar results were found for adsorption on Spheron 6, but with generally lower levels of adsorption. This was presumably due to direct competition of water for the polar sites. On alumina, adipic acid appears to form a virtually complete monolayer in the "parallel" orientation, but succinic acid does not form a complete layer.

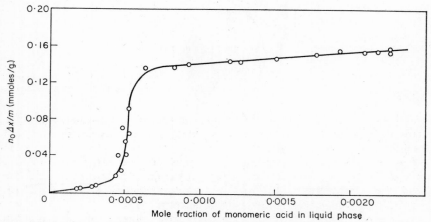

Fig. 7.30. Adsorption of sebacic acid from solutions in di-isopropyl ether on Graphon at 20°c.

The dicarboxylic acids are not very soluble in most simple organic solvents. A few experiments, however, have been carried out with solutions in di-isopropyl ether. Adsorption is generally small, even at high relative concentrations. The only exception to this is sebacic acid, which forms almost a complete monolayer on Graphon. The isotherm shows, however, an unusual sharp rise to this level after remaining very low at low relative concentrations (Fig. 7.30). No explanation of this effect has yet been made.

Phenols

Phenol can be adsorbed from solution either in water or in organic solvents. When it is adsorbed from a non-polar solvent by a polar solid, hydrogen-bonding may occur, e.g. in adsorption from cyclohexane by alumina.[51] The small fraction of the surface of charcoal which is covered at limiting adsorption[49] is probably a measure of the proportion of the surface covered by

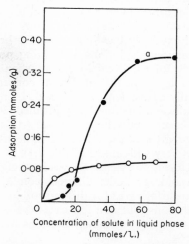

FIG. 7.31. Adsorption on alumina at 58°C from aqueous solutions of (a) phenol, (b) resorcinol.[52] (Reproduced with permission from the *Journal of the Chemical Society*.)

polar (oxygen) groups. Alkyl substituents in the *ortho* position reduce the extent of adsorption, both by a steric effect and by decreasing the acid strength of the phenolic group.

If water is used as the solvent, it competes for the adsorption sites on polar solids, and the isotherm is convex to the concentration axis near the origin (Fig. 7.31a). Resorcinol, with two hydroxyl groups, gives an isotherm (Fig. 7.31b) which is concave to the concentration axis.[52]

On graphite, phenol is readily adsorbed from water.[53] In this case, no interaction with the hydroxyl group can occur, but it is likely that interaction between the π-electron systems of the benzene ring and the graphite occurs, the molecules being adsorbed from solutions of low concentration with the benzene ring parallel to the surface. The isotherm rises in two stages, however, and the second rise is attributed to orientation of the molecules perpendicular to the surface, with the hydroxyl groups pointing away from the graphite surface and towards the polar aqueous phase.

Giles has used *p*-nitrophenol as a solute for determining specific surface areas of solids both polar and non-polar (see Chapter 17).

Iodine

The adsorption of iodine from solution has frequently been used in measurements of the surface areas of charcoals and other carbons. It has been assumed that the adsorption of iodine is independent of the presence of oxide complexes on the surface, and therefore gives the true surface area of carbons. For this reason it has been used to follow the changes in surface area brought about when carbons are exposed to oxidizing or reducing gases at various temperatures. This is important in studies of the effects of surface complexes on the adsorption of strong acids and strong bases from aqueous solution.

It has been assumed explicitly that iodine is adsorbed only as the iodine molecule (even from aqueous solutions containing potassium iodide) and only by van der Waals forces; implicitly, that when so adsorbed, iodine forms a complete monolayer on the surface at concentrations of solution which are experimentally convenient. As a result of recent work, it seems unlikely that these results can be completely justified.

For surface area measurements it is necessary to assume a reasonable value for the molecular area of the adsorbed molecule. The values which have been used cover a very wide range (Table 7.1).

This shows the need for more extensive investigation of the type of adsorption which iodine can undergo under different conditions, and for the use of adsorbents of known surface area for this purpose. Most studies have been made with aqueous solutions, but in a few cases organic solvents have been used; some of these, at least, give simpler solutions than those in potassium iodide.

(a) *Aqueous Solutions*

In adsorption from aqueous solutions, potassium iodide is always present; hence some tri-iodide is formed, and three species containing iodine can potentially be adsorbed. Watson and Parkinson have shown that only free iodine is adsorbed by carbon black.[66] It has similarly been shown that I_2 is adsorbed by platinum preferentially to I^- or I_3^- and can cover nearly a monolayer;[67] molecular iodine is here the least soluble component of the solution. If adsorption isotherms are plotted in terms of "total" iodine concentration (I_2 plus I_3^-), the isotherm changes according to iodide concentration. Concordant results are obtained for adsorption on carbon black, however, if the isotherms are plotted in terms of concentration of free molecular iodine.

TABLE 7.1

Effective Area of Iodine Molecule (sq. Å)	Basis of Calculation	Reference
15·6	Effective radius of molecule in vapour phase as given by viscosity measurements	54, 55
21·2	Gram-molecular volume	56
21·7	Limiting adsorption from aqueous solution on carbon blacks of known specific surface area	57
22·1	Not explained	58
23·5	Crystal structure	59, 60
23·7	Limiting adsorption of vapour and from some organic solutions on Graphon	13
27·1	Limiting adsorption of vapour on Graphon	61
27·4	Molecular dimensions	61
30·0	Limiting adsorption from aqueous solution on Graphon	61
33·0	Limiting adsorption from aqueous solution on carbon blacks of known specific surface area	62
35·3	Adsorption from solution in carbon tetrachloride on "preferred positions" of the magnesium oxide surface	63
42	Limiting adsorption from aqueous solution on carbon black of known specific surface area	64
49·2	Atomic domain radius for iodine molecule	65

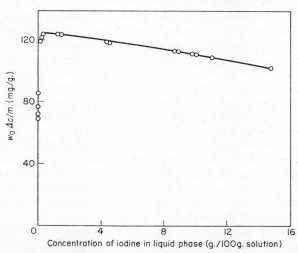

FIG. 7.32. Adsorption of iodine from aqueous potassium iodide by Graphon.

More recently it has appeared that Watson's conclusions may be somewhat over-simplified. For adsorption on Graphon, the isotherm shown in Fig. 7.32 is obtained when results are plotted in terms of total iodine concentration, for experiments in which the ratio of total iodine to total iodide was kept constant. The fall in adsorption at high relative concentration is best explained on the assumption that some I_3^- ions are adsorbed, and that with increasing concentration they displace iodine molecules.[68] As the I_3^- ion almost certainly occupies a greater area on the surface than does the iodine molecule, fewer are adsorbed by unit weight of Graphon, and adsorption appears to fall with increasing concentration.

Watson also found that some of the iodine adsorbed by carbon blacks could not be recovered.[61] The unrecoverable iodine was, in part, retained by the carbon black (presumably chemisorbed), and in part reduced to iodide which remained in the solution. This suggested the possible dehydrogenation of carbon by molecular iodine, and also the adsorption of products of hydrolysis:

$$I_2 + H_2O \rightarrow HOI + HI$$
$$\text{carbon} + HOI \rightarrow \text{carbon (OH).I}$$

or, $$\text{carbon} + HOI \rightarrow \text{carbon (O)} + HI.$$

The iodine chemisorbed by sugar charcoal is recovered by evacuation at 1200°C. A much greater amount of iodine, however, is converted to hydrogen iodide, and results in a corresponding increase in oxygen content of the charcoal,[69] detectable as carbon dioxide when the charcoal is evacuated at 1200°C.

(b) Adsorption from Solution in Organic Solvents

Many of these solutions are simpler than the aqueous solutions, but the variety of colour observed raises the question as to whether adsorption is affected by the state of the iodine in the various solutions.

It has been apparent for many years that the solvent can compete effectively for the surface.[70, 71] In adsorption from carbon tetrachloride, it gave a surface area for glass spheres less than 40% of that determined microscopically;[65] the percentage would be much smaller if a more modern value were used for the area of the iodine molecule. Very low adsorption has also been found on alumina and silica.[68]

A more detailed study has shown that competition from the solvent is general in adsorption on the carbon black, Spheron 6, the extent varying with the solvent. It is much less in adsorption on Graphon, which is almost free from surface complexes, and from some solvents adsorption tends to a complete monolayer.[68] The unusual shape of the isotherm which is sometimes observed is referred to below (p. 131).

Some irreversible adsorption was found to occur on Spheron 6, but no desorption products, such as hydrogen iodide, were found. Chemisorption probably occurs, but probably not altogether by the same mechanisms as were suggested by Watson for chemisorption from aqueous solution. Removal of oxygen and hydrogen complexes by heating the carbon black reduces the extent of chemisorption.

Adsorption of iodine on magnesium oxide is irreversible,[55] and takes place in "preferred positions".[63] Although it is not clear whether it is adsorbed in molecular form or is chemisorbed, dissociatively, as atoms, in either case an area of 35·33 sq. Å would be occupied by each pair of iodine atoms.

These results show that the area associated with an adsorbed iodine molecule depends on:

(i) whether or not chemisorption occurs,
(ii) the nature of the surface on which it is adsorbed,
(iii) the extent of competition from the solvent, if this is organic,
(iv) the effect of iodide concentration (and hence of tri-iodide concentration) in adsorption from aqueous solutions.

Thus a variation in molecular area is to be expected. If a complete monolayer is adsorbed physically, the two more extreme values given in Table 7.1 can now be eliminated, and a value of about 22–24 sq. Å seems normally the most satisfactory to use. In chemisorption, of course, the value is related to the spacing of the active groups on the solid surface.

Metals

An indication of the very wide range of temperature over which adsorption can be measured is given by some results[72] on adsorption from solutions in nickel at 1475°c. These showed that, on alumina, a monolayer of titanium is adsorbed at a concentration of 0·01% by weight, and of chromium at about 1%.

A CLASSIFICATION OF SHAPES OF ISOTHERMS

Although the majority of isotherms for adsorption of solids from solution have the shape shown in Fig. 7.1 (with, in some cases, a rise at high relative concentrations) several other shapes have been observed. These have been classified by Giles[53] according to the scheme shown in Fig. 7.33. The main classification is based on the initial slope of the isotherm, and the subclassification on the shape at higher concentrations. Some of the isotherms refer to adsorbents (e.g. textile fibres) and adsorbates (e.g. dyestuffs) which are outside the scope of this book.

The L curve is of the so-called Langmuir type. It may be found when there is no strong competition from the solvent for sites on the surface. Another possibility is that, if the adsorbate has linear or planar molecules, the major axis is parallel to the surface.

The S curve is obtained if (i) the solvent is strongly adsorbed, (ii) there is strong inter-molecular attraction within the adsorbed layer, (iii) the adsorbate is mono-functional. The second condition is most likely to obtain if the major axis of the adsorbed molecules is perpendicular to the surface. By "mono-functional" is meant, in this context, that the molecule has a single point of strong attachment in an aromatic system, or an aliphatic

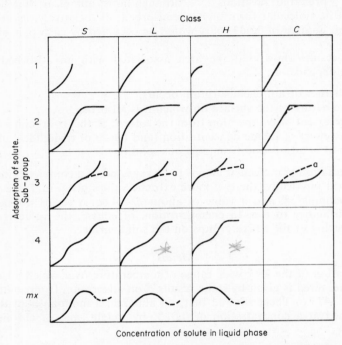

FIG. 7.33. System of classification of isotherms.[53] (Reproduced with permission from the *Journal of the Chemical Society*.)

system of more than five carbon atoms; the adsorbate is, moreover, not micellar. An example is a monohydric phenol, especially if adsorbed on a polar substrate from a polar solvent.

The H curve occurs when there is high affinity between the adsorbate and adsorbent which is shown even in very dilute solutions. Thus it can result from chemisorption or from the adsorption of polymers or ionic micelles, though other special cases are known.

The C curve indicates constant partition of the adsorbate between the solution and the adsorbent. It appears mainly with textile fibres, into which the solute penetrates to further extents as its concentration in the solution is increased.

A number of isotherms show steps which have been discussed above. In general the steps appear to mark a phase-change in the adsorbed layer or the onset of the formation of a second molecular layer after completion of the first. A less common shape involves a wave rather than a step, and is found in the adsorption of iodine from organic solvents by Graphon. An example is shown in Fig. 7.11 for adsorption from cyclohexane. Some other solvents give a more pronounced effect. The change in slope occurs at too small a value of adsorption to be due to the formation of a second layer of molecules, and there is no obvious phase-change which can occur. It is therefore thought that it indicates a form of co-operative adsorption.[13] The iodine molecule is easily deformed, and the packing in the crystalline state indicates some degree of deformation. Similar deformation might occur in the adsorbed film, the lateral interactions increasing the ease of adsorption. This effect could occur only at relatively high surface coverages, and might therefore account for the position of the break between the two halves of the isotherm.

REFERENCES

1. H. Freundlich, "Colloid and Capillary Chemistry", Methuen, London, 1926.
2. E. F. Lundelius, *Kolloid Z.*, 1920, **26**, 145.
3. G. C. Schmidt, *Z. phys. Chem.*, 1910, **74**, 689.
4. J. Guastalla, *Compt. rend.*, 1949, **228**, 820.
5. J. T. Davies, *Trans. Faraday Soc.*, 1952, **48**, 1052.
6. J. T. Davies and E. K. Rideal, "Interfacial Phenomena", Academic Press, New York and London, 1961, Chapter 4.
7. D. A. Haydon, *Kolloid Z.*, 1961, **179**, 72.
8. N. Z. Frangiskos, C. C. Harris, and A. Jowett, "Proceedings of the Third International Congress of Surface Activity", Verlag der Universitätsdruckerei, Mainz, 1961, Vol. 4, p. 404.
9. J. J. Kipling and E. H. M. Wright, *J. Chem. Soc.*, 1962, 855.
10. S. G. Daniel, *Trans. Faraday Soc.*, 1951, **47**, 1345.
11. P. V. Cornford, J. J. Kipling, and E. H. M. Wright, *Trans. Faraday Soc.*, 1962, **58**, 74.
12. R. N. Smith, C. F. Geiger, and C. Pierce, *J. Phys. Chem.*, 1953, **57**, 382.
13. J. J. Kipling, J. N. Sherwood, and P. V. Shooter, *Trans. Faraday Soc.*, 1964, **60**, 401.
14. H. A. Smith and J. F. Fuzek, *J. Amer. Chem. Soc.*, 1946, **68**, 229.
15. E. L. Cook and N. Hackerman, *J. Phys. Chem.*, 1951, **55**, 549.
16. D. M. Brower, *Chem. and Ind.*, 1961, 177.
17. J. J. Rooney and R. C. Pink, *Proc. Chem. Soc.*, 1961, 70; J. J. Rooney and R. C. Pink, *Trans. Faraday Soc.*, 1962, **58**, 1632.
18. C. H. Giles and R. B. McKay, *J. Chem. Soc.*, 1961, 58.
19. J. J. Kipling and E. H. M. Wright, *J. Chem. Soc.*, 1963, 3382.
20. J. J. Kipling and E. H. M. Wright, *J. Chem. Soc.*, 1964, 3535.
21. W. D. Harkins and D. M. Gans, *J. Amer. Chem. Soc.*, 1931, **53**, 2804, cf. also *J. Phys. Chem.*, 1932, **36**, 86.
22. A. S. Russell and C. N. Cochran, *Ind. Eng. Chem.*, 1950, **42**, 1332.
23. A. C. Zettlemoyer, J. J. Chessick, and C. M. Hollabaugh, *J. Phys. Chem.*, 1958, **62**, 489.

24. E. Davidson and J. J. Kipling, unpublished results.
25. D. J. Crisp, *J. Colloid Sci.*, 1956, **11**, 356.
26. J. W. McBain and R. C. Dunn, *J. Colloid Sci.*, 1948, **3**, 303.
27. L. G. Ganichenko, V. F. Kiselev, and K. G. Krasil'nikov, *Proc. Acad. Sci. U.S.S.R.*, 1959, **125**, 1277.
28. R. J. Ruch and L. S. Bartell, *J. Phys. Chem.*, 1960, **64**, 513.
29. G. Venturello, G. Saini, and A. Burdese, *Gazzetta*, 1948, **78**, 254.
30. G. Venturello and A. Burdese, *Gazzetta*, 1948, **78**, 271.
31. G. Venturello and A. M. Ghe, *Gazzetta*, 1959, **89**, 1181.
32. G. Venturello and A. M. Ghe, *Gazzetta*, 1959, **89**, 1191; 1960, **90**, 1266.
33. H. A. Smith and K. A. Allen, *J. Phys. Chem.*, 1954, **58**, 449.
34. H. A. Smith and R. M. McGill, *J. Phys. Chem.*, 1957, **61**, 1025.
35. H. A. Smith and T. Fort, *J. Phys. Chem.*, 1958, **62**, 519.
36. E. D. Tingle, *Trans. Faraday Soc.*, 1950, **46**, 93.
37. F. P. Bowden and A. C. Moore, *Trans. Faraday Soc.*, 1951, **47**, 900.
38. W. Hirst and J. K. Lancaster, *Trans. Faraday Soc.*, 1951, **47**, 315.
39. J. K. Lancaster, *Trans. Faraday Soc.*, 1953, **49**, 1090.
40. N. Hackerman and A. H. Roebuck, *Ind. Eng. Chem.*, 1954, **46**, 1481.
41. S. Claesson, *Rec. Trav. chim.*, 1946, **65**, 571.
42. J. H. de Boer, G. M. M. Houben, B. C. Lippens, W. H. Meijs, and W. K. A. Walrave, *J. Catalysis*, 1962, **1**, 1.
43. J. H. de Boer, M. E. A. Hermans, and J. M. Vleeskens, *Proc. k. ned. Akad. Wetenschap.*, 1957, *B* **60**, 54.
44. N. K. Adam, "The Physics and Chemistry of Surfaces", Clarendon Press, Oxford, 1941, 3rd edition.
45. J. J. Kipling and E. H. M. Wright, unpublished results.
46. E. F. Hare and W. A. Zisman, *J. Phys. Chem.*, 1955, **59**, 335.
47. W. C. Bigelow, D. L. Pickett, and W. A. Zisman, *J. Colloid Sci.*, 1946, **1**, 513, and later papers; E. G. Shafrin and W. A. Zisman in "Mono-molecular Layers", (H. Sobotka, ed.), American Association for the Advancement of Science, Washington, D.C., 1954, p. 129.
48. L. O. Brockway and J. Karle, *J. Colloid Sci.*, 1947, **2**, 277.
49. O. H. Wheeler and E. M. Levy, *Canad. J. Chem.*, 1959, **37**, 1235.
50. L. Morrison and D. M. Miller, *Canad. J. Chem.*, 1955, **33**, 330.
51. B. Eric, E. V. Goode, and D. A. Ibbotson, *J. Chem. Soc.*, 1960, 55.
52. T. Cummings, H. C. Garven, C. H. Giles, S. M. K. Rahman, J. G. Sneddon, and C. E. Stewart, *J. Chem. Soc.*, 1959, 535.
53. C. H. Giles, T. H. MacEwan, S. N. Nakhwa, and D. Smith, *J. Chem. Soc.*, 1960, 3973.
54. L. H. Reyerson and A. W. Wishart, *J. Phys. Chem.*, 1938, **42**, 678.
55. R. C. Dunn and H. H. Pomeroy, *J. Phys. Chem.*, 1947, **51**, 981.
56. A. C. Zettlemoyer and W. C. Walker, *Ind. Eng. Chem.*, 1947, **39**, 69.
57. H. Funk and F. Rämmele, *Chem. Tech. (Berlin)*, 1954, **6**, 213.
58. J. H. de Boer, *Rec. Trav. chim.*, 1946, **65**, 576.
59. P. M. Harris, E. Mack, and F. C. Blake, *J. Amer. Chem. Soc.*, 1928, **50**, 1583.
60. A. I. Kitaigorodskii, T. L. Khotsyanova, and Y. T. Struchkov, *Zhur. fiz. Khim.*, 1953, **27**, 780.
61. J. W. Watson, Ph.D. Thesis, University of London, 1957.

62. W. R. Smith, F. S. Thornhill, and R. R. Bray, *Ind. Eng. Chem.*, 1941, **33**, 1303.
63. W. C. Walker and A. C. Zettlemoyer, *J. Phys. Chem.*, 1953, **57**, 182.
64. M. Smith, Ph.D. Thesis, University of Durham, 1957.
65. A. J. Urbanic and V. R. Damerell, *J. Phys. Chem.*, 1941, **45**, 1245.
66. J. W. Watson and D. Parkinson, *Ind. Eng. Chem.*, 1955, **47**, 1053.
67. W. Lorenz and A. Mühlberg, *Z. phys. Chem. (Frankfurt)*, 1958, **17**, 129.
68. J. J. Kipling and P. V. Shooter, unpublished results.
69. B. R. Puri and D. D. Singh, *J. Indian Chem. Soc.*, 1960, **37**, 401.
70. O. C. M. Davis, *J. Chem. Soc.*, 1907, **91**, 1666.
71. J. B. Firth and F. S. Watson, *J. Chem. Soc.*, 1923, **123**, 1219.
72. C. R. Kurkjian and W. D. Kingery, *J. Phys. Chem.*, 1956, **60**, 961.

CHAPTER 8

The Adsorption of Polymers

INTRODUCTION

The adsorption of polymers is governed by the factors which have been discussed in Chapter 7, but additional features arise which are of considerable importance. The most obvious is that each molecule has a large number of groups, each of which can potentially be adsorbed at the surface. Moreover, the molecule has so many internal degrees of freedom that these groups have a high degree of mutual independence, subject to the important overriding consideration that the adsorption of any group increases the probability of adsorption of other, and especially neighbouring, groups in the molecule.

The polymers which can be studied are almost entirely linear, as any considerable degree of cross-linking reduces solubility to a negligible value. (Adhesion, as opposed to adsorption, remains an important property of cross-linked, as it is of linear polymers.) Even for linear polymers, the absolute solubility limit is low (in most cases rarely exceeding 1 % by weight), but adsorption can often occur so readily that very low concentrations have to be examined. Indeed, the upper limit of investigation may be set, not by the solubility limit (which is often ill-defined), but by the difficulty of working with viscous solutions.

The main features to be investigated in the adsorption have been summarized by Eirich[1] as the effects of molecular weight and dispersity, of temperature, the nature of the solvent and of the adsorbent, and the type of monomer unit in the polymer. The most important matter of interpretation is the configuration of the adsorbed molecules. These topics are considered in the following sections.

A major consideration is that most polymers are polydisperse. In principle, therefore, adsorption from solutions of polymers should in nearly all cases be regarded as adsorption from a multicomponent solution. In practice, however, it has been found very difficult to treat polymer systems in this way. Considerable progress has been made by using fractionated polymers, and it is to be hoped that this will increasingly be the practice in the majority of fundamental studies. Nevertheless, a spread of molecular weight is still found even in carefully fractionated samples. Moreover, the practical use of unfractionated polymers is so great that the problem of marked polydispersity should not be neglected. In investigations on polydisperse systems it is always desirable to ascertain whether adsorption is accompanied by any change in the distribution of molecular weight of the polymer. Viscometry is valuable for this purpose.[2]

The range of available polymers has not yet been investigated systematically, and it remains to be seen whether present results are adequately representative of the general phenomenon. This is probably not the case as far as the choice of adsorbents is concerned. Special attention has been paid to carbons, because it has been hoped that an understanding of their adsorptive properties would throw light on the reinforcing of rubber (both natural and synthetic) by carbon black.

FACTORS AFFECTING ADSORPTION OF POLYMERS

Rate of Adsorption

This factor is important in studying adsorption of polymers because equilibrium may be reached only after several days (or even weeks) instead of the few hours which are usually sufficient for adsorption of substances of low molecular weight. Periods varying from 18 to 290 hours have been found for different combinations of synthetic rubber and carbon black.[3] Moreover, in a polydisperse system, molecules of different molecular weight may not be adsorbed at the same rate.

Because such long periods are involved, some authors refer to the establishment of apparent equilibrium, implying that it is difficult to be certain whether or not true equilibrium has been reached at any given time. The difficulty is emphasized by Fig. 8.1, which refers to the adsorption of polyvinyl acetate from benzene onto chrome plate.[4] Adsorption appears

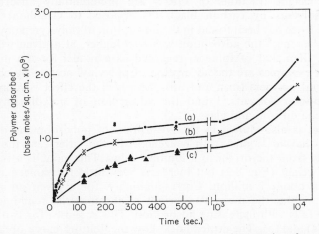

Fig. 8.1. Rate of adsorption of polyvinyl acetate onto chrome plate from solutions in benzene of concentration[4]: A, $1 \cdot 15 \times 10^{-4}$; B, $5 \cdot 75 \times 10^{-5}$; C, $2 \cdot 30 \times 10^{-5}$ moles/litre. (Reproduced with the permission of the American Chemical Society from the *Journal of Physical Chemistry*.)

to be complete after a few minutes, but a plateau in the curve is followed by a further rise which continues up to several hundred hours.

In investigations of this kind, and generally in adsorption of polymers, it is important to recognize that mechanical degradation of polymers (i.e. degradation to lower molecular weight) can occur simply as a result of shaking the solution of polymer with a solid. This has been shown by Sonntag and Jenckel,[5] who have examined in detail the importance of a wide range of conditions.

The rate of adsorption of polymers is discussed further in Chapter 13.

Molecular Weight and Distribution of Molecular Weight

Because polymer samples are rarely as homogeneous or well-defined as the simpler substances which have been considered in previous chapters, results should always be discussed in relation to the information available about the sample of polymer used in a particular experiment. Different samples of any polymer are likely to behave differently in adsorption experiments. Two of the most important variables are molecular weight and distribution of molecular weight.

In some systems, the effect of molecular weight appears to be unimportant. The adsorption of GR–S rubber (a butadiene–styrene copolymer) from benzene by Graphon[6] varied only slightly with molecular weight over the range 32,000 to 230,000. In this system, the sol fraction was appreciably adsorbed, but no adsorption of the "microgel" fraction could be detected.[2]

It is more usual, however, for adsorption to increase with increasing molecular weight. There is some evidence, for example, that the higher molecular weight fractions of GR–S are preferentially adsorbed from aromatic solvents by carbon black (as opposed to Graphon).[7,8] More definite evidence has been found in the adsorption of polystyrene by charcoal from cyclohexane;[9] the adsorption was also higher, at a given weight concentration, than that of toluene, regarded as a model for the monomeric unit. Other examples are the adsorption of polyvinyl acetate by charcoal,[10] and by metal powders (iron and tin), for which the effect is more marked with the poorer solvents,[11] and the adsorption of polydimethylsiloxane from benzene on iron and glass.[12]

The effect is also shown in the adsorption of polyisobutene on carbon black from hexane,[13] and from benzene.[14] In the latter case the isotherm rises steeply from the origin to a limiting value. This value is a function of molecular weight (Fig. 8.2) but increases very little for molecular weights higher than about 1 million. Correspondingly, for a given initial (mean) molecular weight, the molecular weight of the material remaining in the bulk phase falls with increasing adsorption. This is shown for two fractions in Fig. 8.3(a) and (b). The dotted lines show the limiting curve obtained if it is assumed that all material of a given molecular weight is adsorbed before any of the next lower molecular weight, and also that polymers have initially the most probable distribution of molecular weight as given by Flory.[15]

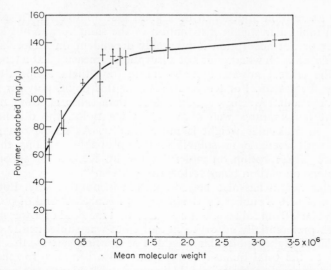

FIG. 8.2. Effect of molecular weight on limiting adsorption of polyisobutene from benzene by carbon black.[14] (Reproduced from the *Journal of Applied Polymer Science*.)

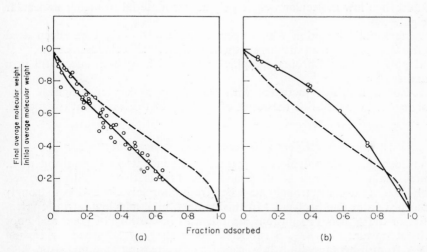

FIG. 8.3. Average molecular weight of polyisobutene remaining in solution after adsorption from benzene by carbon black[14]: (a) viscosity-average molecular weight initially 76,000; (b) viscosity-average molecular weight initially 1,970,000. (Reproduced from the *Journal of Applied Polymer Science*.)

The first assumption appears to be a good first approximation, given that the actual molecular weight distribution of the samples was not known.

Whether or not adsorption is markedly dependent on molecular weight probably depends on whether or not adsorption is measured at a temperature near to the critical miscibility temperature (or "theta temperature", see below) for the system. For temperatures near this point, polymer fractions of higher molecular weight are less soluble than those of lower molecular weight. At temperatures well above the theta point, the dependence of solubility on molecular weight is much smaller. As adsorption generally increases with decrease in solubility, this probably explains the varied dependence of adsorption on molecular weight. Results for adsorption of polyisobutene on carbon black follow this pattern.[16]

Molecular weight has also been important in another way. Both in the adsorption of GR–S rubber by Graphon from benzene,[6] and of polystyrene by carbon black from toluene,[17] it appeared that material of lower molecular weight was preferentially adsorbed initially, but was subsequently displaced by material of higher molecular weight. For the first system, this made no difference to the total quantity of polymer adsorbed, but for the second system, an approximately three-fold increase took place over a period of two weeks. Some care is needed in assessing these results, in view of Sonntag and Jenckel's findings that mechanical degradation can cause a decrease in molecular weight of polymers in the bulk phase. The same result (decrease in molecular weight of unadsorbed polymer) would be observed whether the cause were mechanical degradation of unadsorbed polymer or displacement of adsorbed low molecular weight polymer by material of higher molecular weight.

Differential adsorption with respect to molecular weight has been used to fractionate polymers such as synthetic rubbers,[18,19] cellulose acetate,[20,21] and polystyrene.[22] Chromatographic techniques have been successful in a number of cases.[23,24]

Adsorption which increases with decreasing molecular weight of the polymer is thought to be the result of porosity of the adsorbent (see below, p. 143).

The Effect of Solvent and of Temperature

The adsorption of polymers is markedly dependent on the solvent used. Three main effects can be discerned. The first is that of solvent power. Polymers are usually strongly adsorbed from poor solvents and less strongly adsorbed from good solvents. In this context, high and low intrinsic viscosities of the respective solutions are taken as indications of good and poor solvent power. Thus GR–S rubber is most strongly adsorbed by carbon black from n-heptane, and then in order of decreasing amounts from 90% benzene–10% ethanol, benzene, chloroform, carbon tetrachloride, toluene, and xylene (Fig. 8.4).[6] Similarly the adsorption of polyisobutene on carbon black is much less from cyclohexane or toluene than from benzene.[14]

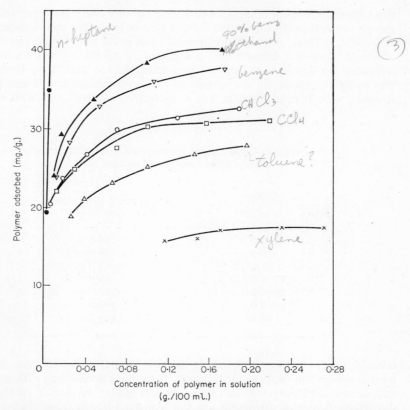

FIG. 8.4. Effect of solvent on adsorption of GR–S rubber by Graphon[6]: ◇, 90% benzene–10% ethanol; ▽, benzene; ○, chloroform; □, carbon tetrachloride; ✕, xylene. (Reproduced with the permission of the American Chemical Society from the *Journal of Physical Chemistry*.)

Even if a limiting value is reached at high relative concentrations for a given solvent, the limiting value may vary markedly as the solvent is changed. An example is given in Table 8.1 for the adsorption of polystyrene by charcoal from methyl ethyl ketone containing small percentages of methanol.[25] The addition of methanol makes the mixture a poorer solvent for the polymer, and increases the limiting adsorption.

There is some evidence that adsorption decreases with increase in the "solubility parameter" of the solvent.[26]

This effect of solvent is not always observed. If very strong interaction occurs between the polymer and the adsorbent, the effect may be very small, as in the adsorption of polylauryl methacrylate on silica from two solvents (n-dodecane and *cis*-decalin) in which it has markedly different intrinsic viscosities.[27] Correspondingly, the second solvent effect may predominate, viz. strong competition between solvent and polymer for the surface of the adsorbent.[28] Thus, although preferential adsorption of polymers is normally observed, "negative adsorption" is known, e.g. in the adsorption of poly-isobutene by Spheron 6 from carbon tetrachloride and from benzene.[16] The oxide complexes on the surface of the carbon black can interact more strongly with benzene (which has a π-electron system) and with carbon

TABLE 8.1

Adsorption of polystyrene by charcoal from mixtures of methyl ethyl ketone and methanol

% MeOH (v/v) in solvent	Limiting adsorption (mg./g.)
0·0	34·0
0·5	35·0
1·0	48·0
1·5	46·0
2·5	81·0
5·0	144·2
10·0	144·2

tetrachloride (which has polarizable chlorine atoms) than with the relatively inert hydrocarbon chain of the polymer. A similar effect occurs with cellulose and solutions of polyvinyl acetate in dioxan and in acetone.[26] In this case the strong forces of attraction are the hydrogen-bonds from the hydroxyl groups of the cellulose chain to the polar centres of the solvents which are probably more accessible than the ester groups of the polymer.

Competition between polymer and solvent for the surface may be evenly balanced and result in there being no apparent adsorption of the polymer, e.g. solutions of polyisobutene in carbon tetrachloride with a particular carbon black,[13] and of hexamethyl disiloxane in benzene with iron and with glass.[12] In such cases a change in solvent is likely to involve both of the above effects.

The third effect is that of temperature, and is somewhat complex. In some systems, decrease in adsorption occurs with increasing temperature, as for most simple solutions. An example is the adsorption of polyisobutene from benzene by carbon black (Fig. 8.5).[14] Other systems, however, show an increase in adsorption with rise in temperature, e.g. the adsorption of polyvinyl acetate by metal powders.[11] This has been attributed to the higher concentrations which can be achieved with a given solvent on increasing the temperature.

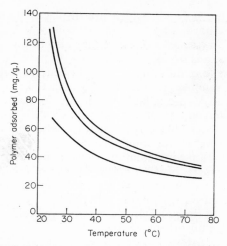

FIG. 8.5. Effect of temperature on adsorption of polyisobutene from benzene by carbon black.[14] The curves (top and bottom) are for polymers of initial average molecular weights of 1,970,000, 1,160,000, and 75,000. (Reproduced from the *Journal of Applied Polymer Science*.)

According to Flory's theory of polymer solutions,[15] the excess free energy of mixing of a polymer and a solvent, ΔG_m^E, is a function of temperature:

$$\Delta G_m^E \propto \left(1 - \frac{\theta}{T}\right), \tag{8.1}$$

where T is the temperature of the solution and θ is the "theta temperature". The latter is the temperature at which deviations from ideality vanish; it can be regarded as the critical miscibility temperature for polymers of infinite molecular weight. The solubility is related to ΔG_m^E which increases markedly with the temperature in the region of θ, but less markedly if T is much higher than θ. Possible effects can be deduced from the theta temperature for polyisobutene which range from 0°A. for n-heptane to 297°A. for benzene.[15] Correspondingly, adsorption of polyisobutene on carbon (at room temperature) is markedly temperature-dependent if the solvent is benzene, but not measurably dependent on temperature if the solvent is n-heptane.[16]

Increasing adsorption with rise in temperature means that the process is endothermic. Hence, for the free energy of the system to decrease, the change in entropy must be positive. Eirich suggests that this occurs because solvent molecules are displaced from the solid surface by polymer molecules and gain more translational entropy than is lost by the polymer molecules. Other possibilities, which are not mutually exclusive, are that the solvent molecules which gain entropy are some of those forming part of the solvation layer surrounding the polymer, which are displaced when the polymer is adsorbed, or are excluded from the polymer coils which may become compressed on adsorption (see below). There may also be an entropy of dilution for the bulk phase as adsorption occurs.

For some systems (e.g. polyvinyl acetate in benzene with copper) a temperature of minimum adsorption is found within the range of temperature normally investigated,[29] presumably when the former of the above two effects becomes more important than the latter with rise in temperature. A broad minimum of this kind may account for adsorption being apparently independent of temperature for some systems.[28]

Polyelectrolytes in aqueous solution show a more complex behaviour than uncharged polymers, because a wider range of conditions can readily be brought about. Thus the adsorption of polymethacrylic acid by anatase is markedly dependent on the pH of the solution,[30] changes in which alter the effective charge on the polymer, the tightness of coiling of its molecules, and the charge on the adsorbent.

The Nature of the Adsorbent

Three properties of the adsorbent are important; the chemical nature of its surface, the specific surface area, and the extent of porosity.

As in adsorption from other systems, the nature of the adsorbing surface determines the extent of competition between solvent and solute and also the extent of adsorption from a given solvent. Thus there is a pronounced variation in the extent of adsorption of GR–S rubber by various commercial carbon blacks,[31] which is not directly related to the specific surface area. The same order appears whether the solvent is n-hexane or chloroform. As the adsorption is increased by heating the carbon blacks to 500°c or 1000°c before use (thus removing some of the oxygen complexes present on the surface) it is probable that the variation is due to the effect of different complexes (or different surface concentrations of the complexes) on the competition between polymer and solvent for the surface. In the same way, polyisobutene is less strongly adsorbed from benzene by heavily oxygenated carbon blacks than by those less heavily oxygenated[14] (Fig. 8.6),[16] and is not adsorbed preferentially by silica.[14]

The specific surface area is important in assessing the extent of adsorption (cf. p. 145). This may include adsorption due to chemical reaction. Thus the amount of the GR–S rubber which reacts chemically with carbon blacks is proportional to the specific surface area for a range of commercial carbon blacks. It is, however, lower for Graphon than would be expected on this basis.[31] This emphasizes that adsorption by unit weight of solid is proportional to specific surface area only for solids with surfaces having the same chemical nature.

If the adsorbent is significantly porous, the specific surface area measured with a small molecule (e.g. nitrogen) may be much better than that accessible

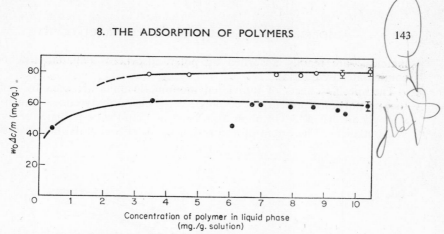

FIG. 8.6. Adsorption of polyisobutene (high molecular weight fraction) from di-isobutene by Spheron 6 (●) and by Graphon (○) at 20°c.

to a polymer molecule. This is suggested as one reason for adsorption being higher on metal foils than on powders.[29]

The effect of porosity of the adsorbent is not well established, because the porous adsorbents which have been used in experiments with polymers have not usually been well characterized. It has been suggested, however, that polymer fractions of low molecular weight may be able to penetrate further into the pore structure of a porous adsorbent than can fractions of higher molecular weight. This would explain the increase in extent of adsorption on charcoal with decreasing molecular weight of polystyrene dissolved in methyl ethyl ketone,[25] of nitrocellulose in acetone, and of dextran in water,[32] and a similar effect in the adsorption of polyvinyl acetate by alumina.[11] The former results have been criticized, however, on the grounds that measurements of adsorption were made before true equilibrium had been established;[17] consequently the true explanation is still uncertain. In the adsorption of polymethacrylic acid from aqueous alcohol on porous charcoal, the extent of adsorption, a, decreased with increasing molecular weight, M, according to the equation:

$$a = AM^{-\beta}, \tag{8.2}$$

where A and β were constants.[33] The decrease in adsorption is thought to be due to exclusion of the larger macromolecular coils from the small pores.

More specifically, it has been found[34] that polydimethylsiloxane is adsorbed from solution in hexane only by the external surface of the molecular sieve 13 X. More experiments of this kind would be valuable in establishing the behaviour of polymers in contact with porous adsorbents.

Chemical Nature of the Polymer

The chemical nature of the polymer in relation to the adsorbent and the solvent has the same significance as in other systems. Thus, while the relatively non-polar polymers are adsorbed readily by carbon adsorbents, polar materials are readily adsorbed by polar adsorbents (e.g. ethylene oxide condensates by quartz).

The specific effects of individual monomeric units are seen by varying the proportions present in co-polymers. Thus polystyrene is much less strongly adsorbed by Graphon from benzene than the co-polymer of styrene and butadiene. The extent of adsorption varies only slightly, however, when

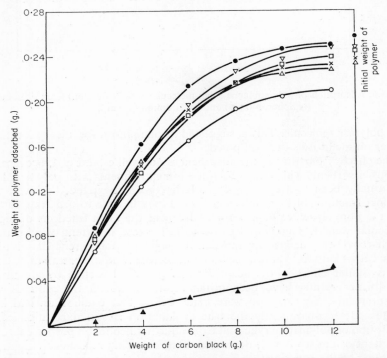

Fig. 8.7. Adsorption of a co-polymer of styrene and butadiene from solution in benzene by Graphon as a function of co-polymer composition[6]: △, 10% butadiene, 90% styrene; ▽, 75% butadiene, 25% styrene; □, 50% butadiene, 50% styrene; ✕, 25% butadiene, 75% styrene; △, 20% butadiene, 80% styrene; ○, 100% butadiene, 0% styrene; ▽, 0% butadiene, 100% styrene. (Reproduced with the permission of the American Chemical Society from the *Journal of Physical Chemistry*.)

the co-polymer contains more than 10% of butadiene (Fig. 8.7).[6] Partially hydrolysed polyvinyl acetate is more strongly adsorbed from dichloroethane by iron than the unhydrolysed polymer.[11] This is clearly due, in the first instance, to strong interaction between the hydroxyl groups and the oxide coating on the metal. As a secondary effect, the dichloroethane would be a somewhat poorer solvent for the alcohol–acetate co-polymer than for the original homopolymer.

THE NATURE OF THE ADSORBED PHASE

For most systems the adsorption isotherm either reaches a marked plateau or tends towards a limiting value; S-shaped curves are not generally observed. It therefore seems unlikely that multilayer formation occurs, at least of the kind described for simpler systems. The absence of a point of inflexion is not considered by all authors, however, to exclude the possibility of multi-layer formation.[35]

The limiting weight of polymer adsorbed may, nevertheless, be many times greater than that of an analogous compound with a simple molecule, under comparable conditions. For example, the adsorption of polyvinyl acetate from carbon tetrachloride on copper is stated to be 20–40 times that of ethyl acetate (regarded as a monomeric analogue of the repeating unit).[11] The significance of the limiting value is thus of great importance in understanding the way in which polymers are adsorbed.

A simple model of an adsorbed (linear) polymer molecule is provided by extrapolation from the long-chain hydrocarbons described in the previous chapter, i.e. the polymer molecule is regarded as a fully-extended linear chain lying on the surface. On this basis a monolayer value can be calculated for any non-porous adsorbent of known specific surface area. If the polymer molecule has large side-groups, more than one configuration may have to be considered.

In a few cases, the limiting value of the isotherm is equal to or is less than this monolayer value. Thus the adsorption of polyesters from chloroform on silica corresponds to a single layer and on alumina to 0·2 of a layer.[36] More usually, however, the limiting value corresponds to several molecular thicknesses. Thus the polyesters apparently formed 2·2 molecular layers when adsorbed from chloroform on glass and 5 layers when adsorbed from toluene. Evidence of thick layers has been obtained from viscometric studies. Polystyrene has been shown, in this way, to form an adsorbed layer 1000–1500 Å thick on the walls of a capillary viscometer.[37,38] The viscosity of suspensions of carbon black in solutions of GR–S rubber in xylene corresponds to that of particles having a coating of rubber 150–200 Å thick.[39] Values of several thousand Å have been considered by other authors.[35]

Such thicknesses are compatible with the existence of many molecular layers in the adsorbate, but the latter concept is not immediately acceptable. There are, for example, no breaks in the adsorption isotherms which might correspond to the completion of the first or of any other layer. Such breaks are now becoming familiar in at least some instances of multilayer adsorption for other kinds of system. Alternative models for the adsorbed molecules have therefore been considered.

An early suggestion was due to Jenckel and Rumbach.[40] They considered the possibility that the adsorbed molecules streamed out from the solid surface into the liquid phase, being attached at one or only a few points per molecule (Fig. 8.8). These points of attachment are now referred to as the "anchor segments" of the polymer chain. A distinction can thus be made between "totally-deposited" and "partially-deposited" polymer molecules. In the latter arrangement much more material can be attached to a given area of surface than in the former.

This type of model has been developed by considering the configuration of the polymer molecules in solution.[15] Molecules of high molecular weight are thought to form random coils which are approximately spherical when

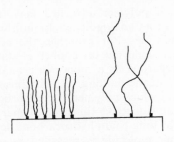

FIG. 8.8. Suggested configurations for adsorbed polymer molecules.[40] (Reproduced with the permission of Verlag Chemie from *Zeitschrift für Elektrochemie*.)

the molecular weight is sufficiently high. These coils occupy a large volume when the polymer is dissolved in a "good" solvent, but a much smaller volume in a "poor" solvent. The size is also dependent on the rigidity of the polymer chain (a property associated with the nature of the monomer unit) and on the strength of interaction between units in the chain. For dilute solutions, the individual coils tend to form separate domains from which units of neighbouring molecules are excluded, but entanglement may occur in concentrated solutions.

It thus seems possible that the polymer molecules might be adsorbed in the form of coils. In this configuration, relatively few monomer units would function as anchor segments, and the adsorbed layer could have a considerable thickness. If the polymer coils were adsorbed in single layers, the maximum adsorption, A_∞, would be given by:

$$A_\infty = \frac{SM}{\pi R^2 N},\qquad(8.3)$$

where A_∞ is in g./g., S is the specific surface area (accessible to the polymer), M is the molecular weight of the polymer, R is the "radius of gyration" of the polymer coil, and N is Avogadro's number. Equation (8.3) was found to give a value of A_∞ several times smaller than the experimental values for adsorption of polyvinyl acetate from various solvents on metal powders.[11] It may therefore be necessary to assume that considerable compression of the polymer coils occurs on adsorption and that interpenetration (entanglement) of the coils can occur at high surface coverage.

For polymers in solution,

$$R^2 \propto M^\beta, \tag{8.4}$$

where β varies between 1 for a poor solvent and 1·4 for a good solvent. Thus from equation (8.3)

$$A_\infty \propto \frac{1}{M^{\beta-1}}, \quad \text{or} \quad A_\infty \propto M^{1-\beta}, \tag{8.5}$$

so that the degree of adsorption is more sharply dependent on molecular weight if a poor solvent is used than if a good solvent is used. This is found to be the case qualitatively.[11] It has, however, been suggested that $(1-\beta)$ is not a constant, but increases with increasing molecular weight,[35] the dependence also varying with temperature.

If the polymers exist as coils in solution, a large adverse entropy change would be involved if they were adsorbed in a linear configuration. This configuration must be considered, however, as it has been observed at the water–air interface. Similarly up to 40% of the segments of polymethacrylic acid may be adsorbed at the water–mercury interface.[41] The two interfaces should not, however, be regarded as strictly comparable.[12] Water is not a good solvent for organic polymers, and a coil of polymer is not likely to extend into the liquid phase. At the liquid–solid interfaces such extension is possible when adsorption occurs from solution.

It is probably useful to regard the two configurations described above as limiting cases, between which a variety of situations can exist. Thus infrared adsorption studies have shown that 35 to 50% of the carbonyl groups of polylauryl methacrylate may be adsorbed at a silica surface when dodecane is used as the solvent.[27] In accordance with this, sedimentation studies show that the adsorbed layer is about 25 Å thick. Strong interaction between polymer and surface is likely to favour total deposition, and weak interaction to favour partial deposition (i.e. adsorption as coils). A model in which this effect can be treated theoretically has been given by Higuchi.[42]

The configuration adopted may depend on the degree of surface coverage. Gottlieb has suggested that at low surface coverage, polyvinyl acetate is adsorbed (from various solvents) onto chromium-plated steel with all of the acetate groups in contact with the surface. At higher degrees of adsorption, the polymer may be adsorbed in the form of coils with a small number of anchor segments, or even as coils attached to a first layer of completely deposited molecules.[4, 43]

EQUATIONS FOR THE ADSORPTION ISOTHERM

The adsorption isotherm for polymers has sometimes been described in terms of the Freundlich equation, though this is not usually satisfactory because the isotherm normally reaches a limiting value.[9] Consequently, an equation of the Langmuir type has been used,[25] e.g.

$$\frac{a}{a_\infty} = \frac{bc}{1+bc},$$ (8.6)

where a is the amount adsorbed at concentration c, and a_∞ is the limiting adsorption, b being a constant.

In view of the comments made previously about the use of the Langmuir equation the significance of the constants in equation (8.6) must be carefully considered. As a first step, it should be emphasized that the composite isotherm cannot necessarily be taken to be equivalent to the individual isotherm for adsorption of the polymer. This would normally be a very good approximation (Chapter 7) for very dilute solutions. If, however, the adsorbed polymer exists in the form of large coils, then especially if the coils are loose, large numbers of solvent molecules can be associated with each polymer molecule. In the equation:

$$\frac{w_0 \Delta c}{m} = w_1^s (1-c) - w_2^s c,$$ (3.6)

although c may be very small, this may be offset by w_2^s being very large, and the term $w_2^s c$ is therefore not necessarily small in relation to w_1^s. Hence $w_0 \Delta c/m$ should not automatically be equated with w_1^s. This point has frequently been overlooked.

A model for adsorption of flexible polymers from dilute solution has been provided by Simha, Frisch, and Eirich.[44-47] It is assumed that the polymer forms a localized monolayer on the surface of the adsorbent, that it is extremely unlikely that each segment of the polymer chain (which is probably coiled to varying extents when the polymer is in solution) is attached to the surface, and that there is a Gaussian distribution of end-to-end distances in the polymer. The remainder of the polymer chain forms loops or bridges which extend into the solution. Interaction between chains near the surface is neglected. The probability that a given segment of the chain is adsorbed by the surface is small. In the limiting case, in which this probability tends to zero, the following equation is observed:

$$\frac{\theta}{1-\theta} \cdot e^{2K_1\theta} = (Kc)^{1/\langle v \rangle}$$

which approximates to

$$\theta = \frac{(Kc)^{1/\langle v \rangle}}{1+(Kc)^{1/\langle v \rangle}},$$ (8.7)

where θ is the fraction of the surface covered by adsorbed segments at a concentration c, and $\langle v \rangle$ is the average number of segments adsorbed on the surface per chain, K_1 and K being constants. This equation requires a steep rise in the isotherm at low concentrations, after which θ depends only slightly on c, and for a considerable range of c remains below the value which would be given by a Langmuir equation (Fig. 8.9). The equation is derived on the assumption that no solvent is adsorbed, but if adsorption of solvent is allowed for, the form of the equation is not significantly affected.

For a long flexible chain and low surface coverage, $\langle v \rangle$ is proportional to $t^{\frac{1}{2}}$, where t is the number of adsorbable segments in the chain. Thus the

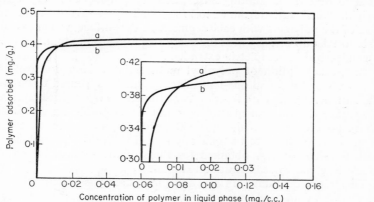

FIG. 8.9. Comparison of isotherms for adsorption of a polymer by equations: (a) of the Langmuir type, (b) by the Simha, Frisch, and Eirich theory.[45] (Reproduced with the permission of the American Chemical Society from the *Journal of Physical Chemistry*.)

extent (and heat) of adsorption should be proportional to the square root of the molecular weight of the polymer. At higher surface coverages, interaction between adsorbed segments becomes important, and adsorption tends to become more nearly a linear function of molecular weight. The value of $\langle v \rangle$ should also increase with increasing flexibility of the polymer chain. The increase of flexibility which results from increase in temperature could outweigh other factors and explain observed cases of increased adsorption with increase in temperature.

An early test of equation (8.7) was made by Kraus and Dugone,[3] who first simplified it. The constant K_1 expresses the interaction of the adsorbed segments in excess of their interaction in bulk. If this is assumed to be zero, the equation becomes:

$$\frac{\theta}{1-\theta} = (Kc)^{1/\langle v \rangle}, \tag{8.8}$$

where θ can be replaced by $\dfrac{a}{a_\infty}$. This was found to give a reasonable representation of the isotherms for adsorption of a number of synthetic rubbers from n-heptane on a number of carbon blacks. The isotherms could not be fitted by an equation of the Langmuir type. In general a good correlation was also found between a_∞ and the surface areas of the carbon blacks. Values of $\langle v \rangle$ ranged from 1·37 to 13·0.

Binford and Gessler found a value of 2·6 for $\langle v \rangle$ in adsorption of butyl rubber of molecular weight 325,000 from hexane on carbon black.[13] This implied such a length for the loops of polymer projecting into the solution that they preferred an alternative interpretation. A Langmuir equation was used to obtain the limiting adsorption for polymers of varying molecular weights, which varied from 62 to 77 mg./g. This would correspond to adsorption of the polymers with each segment on the surface, which appeared to be able to take up one segment for every surface carbon atom, i.e. the polymer was adsorbed in completely extended form.

In the adsorption of polyalkyl methacrylates from dodecane by silica, between one-third and one-half of the segments appeared to be attached to the surface.[27] This high fraction is attributed to hydrogen-bonding between the carbonyl groups of the polymer and —OH groups on the silica surface. Corresponding to the high percentage of anchor segments, the adsorbed layer was calculated to be relatively thin (20–40 Å). This may be compared with a value of 25 Å obtained for films of the same polymer on carbon black, as given by sedimentation studies.

This test of the Simha, Frisch, and Eirich theory is not entirely satisfactory because the data used were for high surface coverages, whereas the equation was derived initially for low surface coverage. Gilliland and Gutoff[40] and Higuchi[42] have criticized the above treatment and have put forward modified equations which apply to adsorption at the theta temperature.

An alternative approach is to consider a kinetic picture in which the following equilibria exist:

1 adsorbed polymer molecule, with v anchor segments
$\rightleftharpoons v$ "free" sites + 1 polymer molecule in solution,

1 adsorbed solvent molecule $\rightleftharpoons$ 1 "free site" + 1 solvent molecule in solution.

Application of the Law of Mass Action leads to the equation:

$$\frac{\theta}{v(1-\theta)^v} = Kc. \qquad (8.9)$$

For the adsorption of polystyrene from toluene by charcoal, equation (8.9) was found to fit the isotherm with $v = 50$, which was obtained as the square root of the degree of polymerization of the polymer.[17] (When $v = 1$, equation [8.9] has the Langmuir form.) Equation (8.9) fits the above data significantly better than an equation of the Langmuir form (Fig. 8.10). In general, however, it appears to be difficult to distinguish between the effectiveness of the two equations. Thus Fontana and Thomas[27] claimed that their isotherms could be fitted satisfactorily by an equation of the Langmuir form for adsorption of polymers with $v \sim 470$ and ~ 1660, respectively.

The above equations all have one important limitation—the assumption that v, the number of anchor segments per chain, is independent of θ, the surface coverage. This is most unlikely to be the case,[42] and more adequate equations for the adsorption isotherm should take account of the variation.

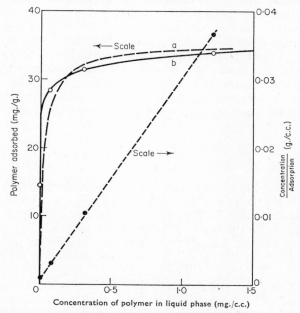

FIG. 8.10. Adsorption of polystyrene from solution in toluene on carbon,[17] at 25°C. (Reproduced from the *Journal of Polymer Science*.)

Silberberg has put forward models for the adsorption process which relate the number of segments in an adsorbed loop to the average energy of adsorption per surface site.[49] He concludes that a high percentage of segments are held at the surface for relatively low adsorption energies (about one kT per segment). A more detailed treatment is given in a subsequent paper.[50]

The possibility of multilayer adsorption of polymers is described by a treatment analogous to that of Brunauer, Emmett, and Teller for a single gas. It is argued that the use of this treatment is justified by the assumption that the solvent is not adsorbed. In addition to the assumptions of the B.E.T. theory,[51] it is assumed that the average number of adsorbed segments is the average of the values of $\langle v \rangle$ for each adsorbed layer. The equation for the isotherm is then:

$$y = \frac{tkv_m c^{1/\langle v \rangle}}{\langle v \rangle (c_0^{1/\langle v \rangle} - c^{1/\langle v \rangle})[1 + (k-1)(c/c_0)^{1/\langle v \rangle}]}, \tag{8.10}$$

where y is the number of polymer chains removed from solution by unit weight of solid, t is the number of segments in a polymer chain, v_m is the number of segments needed to cover the surface of unit weight of solid with a unimolecular layer, c_0 is the limiting concentration of polymer at a given temperature, and k is a constant. The form of this equation is very similar to that of the original B.E.T. equation.

An equation of state has been proposed for polymers adsorbed at interfaces.[52] Although it has not been applied as yet to adsorption at the liquid–solid interface, this development may occur. A number of limiting assumptions are made. The model is a two-dimensional quasi-lattice, energies of interaction are neglected, and it is assumed that, on adding a polymer molecule to the lattice, the segments of those molecules already placed in the lattice are randomly situated among occupied sites. The film pressure is given by:

$$\frac{\pi}{\pi_0} = \left[\frac{(x-1)z'}{2x} \cdot \ln\left(1 - \frac{2A_0}{z'A}\right) - \ln\left(1 - \frac{A_0}{A}\right) \right], \qquad (8.11)$$

where x is the number of monomer units in the molecule (each occupying the same surface area as a molecule of solvent), z' is a constant related to the co-ordination number for the lattice, A is the area occupied per molecule at pressure π, A_0 is the limiting area of the molecule in a condensed film, and $\pi_0 = kTx/A_0$. The equation has been applied to insoluble films rather than to "soluble" films. It has been developed by Frisch and Simha.[53]

The case of rigid (or rod-shaped) solute molecules has been considered by Mackor and van der Waals.[54] They show that, if allowance is made for interaction between adsorbed molecules due to steric hindrance, the isotherm lies much lower than the corresponding "Langmuir" isotherms. The slope of the isotherm becomes very small at surface coverages much below 1.

DESORPTION

It is usually difficult to desorb polymers from a solid surface. For example, of the polyvinyl alcohol adsorbed onto a silver bromide sol from aqueous solution, 1–1·5% was removed by shaking with water at 25°c for 45 hours, and only 13% by boiling for 3 hours.[55] Only 10–15% of polystyrene adsorbed on carbon black could be recovered by extraction with boiling toluene in a Soxhlet apparatus.[17]

If the polymer is weakly adsorbed, desorption is most easily effected by use of a "good" solvent. If it is strongly adsorbed, however, it is better to use a desorbing agent which competes strongly for the surface, e.g. acetonitrile for polyvinyl acetate adsorbed on iron.[11] These procedures, however, do not always result in the complete removal of the polymer even if long periods of desorption (up to a month) with hot solvents are used.

A long period of desorption is always likely to be found whenever the adsorption isotherm has a plateau which extends down to very low concentrations. The pure solvent can then take up only a very small amount of polymer before equilibrium is established. The rate of transport of the polymer away from the surface is consequently very low. The plateau is generally maintained to lower concentrations with a poor solvent than with a good solvent.

Very slow desorption of simpler adsorbates would suggest the possibility of their being chemisorbed. This sometimes occurs with polymers, but is thought not to occur in the majority of systems which have been investigated. This would be expected from a consideration of the chemical groupings involved. Moreover, the variation in the extent of adsorption according to the solvent used implies that any given anchor-segment of the polymer chain is not strongly held by the surface. The variation in the rate of desorption with different solvents supports this conclusion.

The difficulty of desorption, even for polymers held physically to the surface, thus appears to be due to the statistical improbability of removing all of the anchor-segments from the surface simultaneously. In accordance with this, evidence is accumulating that the amount of polymer which can be desorbed in a given time is a function of molecular weight. For example,[14] the quantities of polyisobutene left on carbon black after extraction with benzene at 25°c for 27 days rose from 0·0135 g./g. for polymer of initial molecular weight 74,000, to 0·080 g./g. for polymer of initial weight 1,950,000.

An interesting example of ready desorption is the removal of vinyl polymers from platinum foil by simple solvents.[29] This has not been explained. It does raise the question, however, as to whether removal from other non-porous solids, which are normally finely-divided, is difficult because a given polymer molecule is simultaneously attached to more than one particle of adsorbent. This would be most unlikely with sheets of foil.

Hysteresis

The difficulty in desorbing polymers leads to hysteresis which can be plotted if measurements are made at the temperature at which adsorption took place. Examples are shown in Fig. 8.11 for synthetic rubber and carbon blacks.[3] In this case hysteresis can be shown to be due mainly to chemical reaction between the rubber and carbon black, giving rise to "bound-rubber", which is formed in increasing amounts as the temperature is raised. It is likely, however, that similar hysteresis curves would be found for systems in which chemical reaction did not occur.

A special case has been noted in adsorption of polydimethylsiloxane on glass.[12] As chain interchange within the polymer is possible, it is likely that a similar process takes place with Si—O—Si groups of the glass surface, resulting in formation of chemisorbed chains and possibly, on repetition of the process, gradual reduction in the molecular weight of the polymer desorbed into solution.

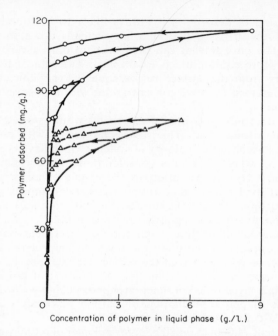

FIG. 8.11. Hysteresis in adsorption of rubber from solution in n-heptane by two types of carbon black.[4] (Reproduced with the permission of the American Chemical Society from *Industrial and Engineering Chemistry*.)

POLYMERS OF LOW MOLECULAR WEIGHT

The above discussions apply to molecules with molecular weights from tens of thousands upwards. Some materials are polymeric in character, but have molecular weights in the region of one thousand or sometimes less. Examples are the polyoxyethylated phenols. These are not strictly polymers, but condensates of a short polymer chain (of ethylene oxide units) with substituted phenols. Their most immediately interesting feature is the balance of hydrophobic and hydrophilic material in the molecule. A comparable condensate is that formed between ethylene oxide and propylene oxide, for which rather higher molecular weights are possible.

The adsorption of ethylene oxide–propylene oxide condensates from aqueous solutions by quartz gives isotherms which can be described by the Langmuir equation. The limiting area for a material of molecular weight 2850 was, at 2880 sq. Å per molecule, so high as to suggest adsorption of a highly solvated molecule.[56]

The adsorption of polyoxyethylated octyl- and nonyl-phenols on quartz is strongly dependent on the proportion of ethylene oxide in the condensate

(Fig. 8.12).[57] Adsorption decreases as the solubility in water is increased
by increasing the proportion of ethylene oxide. The isotherms show very
definite limits to adsorption at low relative concentration.

In similar adsorption of polyoxyethylated nonyl-phenol on calcium
carbonate, adsorption appears to be negligible below a concentration which
is close to the critical micelle concentration for the system (Fig. 8.13).[58]

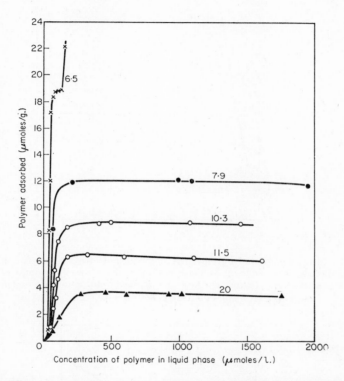

FIG. 8.12. Effect of composition on adsorption of polyoxyethylated nonyl-phenols from
aqueous solution by quartz. The numbers on the isotherm indicate the molar ratio of
ethylene oxide to phenol.[57] (Reproduced with the permission of the American Chemical
Society from the *Journal of Chemical and Engineering Data*.)

This implies that water competes effectively with single molecules of solute
for the surface, and that the isotherms show adsorption of micelles. The
extent of micellar adsorption is even more dependent on the hydrophilic-
lipophilic balance of the condensate than in the adsorption on quartz. In
adsorption on carbon black, however, adsorption occurs at low concentrations and the isotherm can be fitted by a Langmuir equation.[59] Presumably

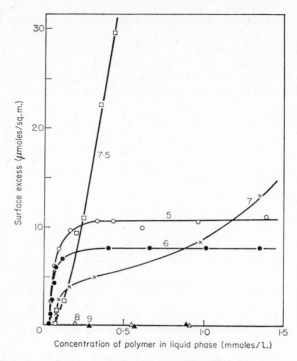

FIG. 8.13. Adsorption of polyoxyethylated nonyl-phenols from aqueous solutions by calcium carbonate. The numbers on the isotherms indicate the molar ratio of ethylene oxide to phenol.[58] (Reproduced from *Kolloid Zeitschrift* with permission of Dr. Dietrich Steinkopff Verlag, Darmstadt.)

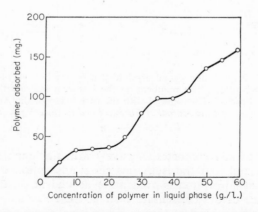

FIG. 8.14. Adsorption of polyethylene glycol (M. W., 6000) from benzene on aluminium foil[61] at 22·5°C. (Reproduced with permission from *Die Makromolekulare Chemie*.)

the difference in behaviour is due to only weak competition from solvent molecules for the surface of carbon black.

It may prove to be relevant that in the adsorption of very low polymers, the free energy of adsorption is directly proportional to the molecular weight. This has been found for the adsorption of glycine and its di-, tri- and tetra-peptides from aqueous solutions by calcium montmorillonite.[60]

Some unusual features have been found in the adsorption of polyethylene glycol from benzene on aluminium and on glass.[61] The isotherm has marked steps (Fig. 8.14). The authors interpret this as being due to multilayer formation. This needs more detailed interpretation, however, as the second "layer" appears to contain more material than the first. Moreover, in adsorp-

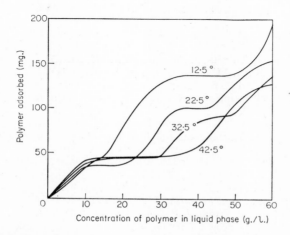

FIG. 8.15. Effect of temperature on adsorption of polyethylene glycol (M. W., 6000) from benzene on aluminium foil.[61] (Reproduced with permission from *Die Makromolekulare Chemie.*)

tion from methanol, the height of the second step is greater than in adsorption from benzene, although the height of the first step is the same. The effect of temperature is also interesting (Fig. 8.15). The height of the first step is almost constant, but the height of the second step is markedly dependent on temperature.

The height of the first step increases with molecular weight of the polymer, and is proportional approximately to the square root of the molecular weight. This power ($\frac{1}{2}$) lies between the values expected for incompressible coils (0) and for fibres, presumably perpendicular to the surface (1). Thus the adsorbed molecules can be regarded as extended loops or very deformed coils.

REFERENCES

1. R. Ullman, J. Koral, and F. R. Eirich, "Proceedings of the Second International Congress of Surface Activity", Butterworths, London, 1957, Vol. III, p. 485.
2. I. M. Kolthoff and A. Kahn, *J. Phys. Chem.*, 1950, **54**, 251.
3. G. Kraus and J. Dugone, *Ind. Eng. Chem.*, 1955, **47**, 1809.
4. C. Peterson and T. K. Kwei, *J. Phys. Chem.*, 1961, **65**, 1330.
5. F. Sonntag and E. Jenckel, *Kolloid Z.*, 1954, **135**, 1, 81.
6. I. M. Kolthoff, R. E. Gutmacher, and A. Kahn, *J. Phys. Chem.*, 1951, **55**, 1240.
7. L. E. Amborski, C. E. Black, and G. Goldfinger, *Rubber Chem. Technol.*, 1950, **23**, 417.
8. G. Goldfinger, *Rubber Chem. Technol.*, 1945, **28**, 286.
9. E. Treiber, G. Porod, W. Gierlinger, and J. Schurz, *Makromol. Chem.*, 1953, **9**, 241.
10. I. Claesson and S. Claesson, *Arkiv Kemi*, 1945, **19**, No. 5.
11. J. Koral, R. Ullman, and F. R. Eirich, *J. Phys. Chem.*, 1958, **62**, 541.
12. R. Perkel and R. Ullman, *J. Polymer Sci.*, 1961, **54**, 127.
13. J. S. Binford and A. M. Gessler, *J. Phys. Chem.*, 1959, **63**, 1376.
14. E. R. Gilliland and E. B. Gutoff, *J. Appl. Polymer Sci.*, 1960, **3**, 26.
15. P. J. Flory, "Principles of Polymer Chemistry", Cornell University Press, Ithaca, N.Y., 1953.
16. E. Davidson and J. J. Kipling, unpublished results.
17. H. L. Frisch, M. Y. Hellman, and J. L. Lundberg, *J. Polymer Sci.*, 1959, **38**, 441.
18. I. Lander, *Compt. rend.*, 1947, **225**, 629; *Rubber Chem. Technol.*, 1948, **21**, 682.
19. M. A. Golub, *J. Polymer Sci.*, 1953, **11**, 583.
20. B. Miller and E. Pacsu, *J. Polymer Sci.*, 1959, **41**, 97.
21. H. Mark and G. Saito, *Monatsh.*, 1936, **68**, 237.
22. S. J. Yeh and H. L. Frisch, *J. Polymer Sci.*, 1958, **27**, 149.
23. G. M. Guzman, "Progress in High Polymers", Heywood, London, 1961, Vol. 1, p. 113.
24. S. R. Caplan, *Lectures, Monographs and Reports*, Royal Institute of Chemistry, London, 1954, No. 5, p. 16.
25. J. F. Hobden and H. H. G. Jellinek, *J. Polymer Sci.*, 1953, **11**, 365.
26. J. E. Luce and A. A. Robertson, *J. Polymer Sci.*, 1961, **51**, 317.
27. B. J. Fontana and J. R. Thomas, *J. Phys. Chem.*, 1961, **65**, 480.
28. S. Ellerstein and R. Ullman, *J. Polymer Sci.*, 1961, **55**, 123.
29. F. Patat and C. Schliebener, *Makromol. Chem.*, 1961, **44–46**, 643.
30. G. Lopatin and F. R. Eirich, "Proceedings of the Third International Congress of Surface Activity", 1960, Verlag der Universitätsdruckerei Mainz, Vol. II, p. 97.
31. I. M. Kolthoff and R. G. Gutmacher, *J. Phys. Chem.*, 1952, **56**, 740.
32. S. Claesson, *Discuss. Faraday Soc.*, 1949, **7**, 321.
33. L. Huppenthal, *Roczniki Chem.*, 1963, **37**, 1001.
34. A. V. Kiselev, V. N. Novikova, and Y. A. El'tekov, *Proc. Acad. Sci. (U.S.S.R.)*, 1963, **149**, 210.
35. F. Patat, E. Killmann, and C. Schliebener, *Makromol. Chem.*, 1961, **49**, 200.

36. R. R. Stromberg, A. R. Quasius, S. D. Toner, and M. S. Parker, *J. Res. Nat. Bur. Stand.*, 1959, **62**, 71.
37. O. E. Öhrn, *J. Polymer Sci.*, 1955, **17**, 137; cf. R. F. Boyer and D. J. Streeter, *J. Polymer Sci.*, 1955, **17**, 154.
38. H. G. Fendler, H. Rohleder, and H. A. Stuart, *Makromol. Chem.*, 1956, **18–19**, 383.
39. L. E. Amborski and G. Goldfinger, *Rubber Chem. Technol.*, 1950, **4**, 803.
40. E. Jenckel and B. Rumbach, *Z. Elektrochem.*, 1951, **55**, 612.
41. I. R. Miller, *Trans. Faraday Soc.*, 1961, **57**, 301.
42. W. I. Higuchi, *J. Phys. Chem.*, 1961, **65**, 487.
43. M. H. Gottlieb, *J. Phys. Chem.*, 1960, **64**, 427.
44. R. Simha, H. L. Frisch, and F. R. Eirich, *J. Phys. Chem.*, 1953, **57**, 584.
45. H. L. Frisch and R. Simha, *J. Phys. Chem.*, 1954, **58**, 507.
46. H. L. Frisch, *J. Phys. Chem.*, 1955, **59**, 633.
47. H. L. Frisch and R. Simha, *J. Chem. Phys.*, 1957, **27**, 502.
48. E. R. Gilliland and E. B. Gutoff, *J. Phys. Chem.*, 1960, **64**, 407.
49. A. Silberberg, *J. Phys. Chem.*, 1962, **66**, 1872.
50. A. Silberberg, *J. Phys. Chem.*, 1962, **66**, 1884.
51. S. Brunauer, "Physical Adsorption of Gases and Vapours", Oxford University Press, London, 1944.
52. S. J. Singer, *J. Chem. Phys.*, 1948, **16**, 872.
53. H. L. Frisch and R. Simha, *J. Chem. Phys.*, 1956, **24**, 652.
54. E. L. Mackor and J. H. van der Waals, *J. Colloid Sci.*, 1952, **7**, 535.
55. S. E. Sheppard, A. S. O'Brien, and G. L. Beyer, *J. Colloid Sci.*, 1946, **1**, 213.
56. H. R. Heydegger and H. N. Dunning, *J. Phys. Chem.*, 1959, **63**, 1613.
57. H. N. Dunning, *Ind. Eng. Chem., Chem. Eng. Data Series*, 1957, **2**, 88.
58. H. Kuno and R. Abe, *Kolloid Z.*, 1961, **177**, 40.
59. R. Abe and H. Kuno, *Kolloid Z.*, 1962, **181**, 70.
60. D. J. Greenland, R. H. Laby, and J. P. Quirk, *Trans. Faraday Soc.*, 1962, **58**, 829.
61. E. Killmann and G. Schneider, *Makromol. Chem.*, 1962, **57**, 212.

CHAPTER 9

Adsorption in Multicomponent Systems

Most fundamental studies of adsorption from solution have been confined to two-component liquids and a single solid, as described in the preceding chapters. The practical use of adsorption, however, is normally that of separating a mixture of solutes present in a common solvent, either by the batch processes which have been used for a very long time or, in more recent years, by chromatography.

Except in relation to chromatography (Chapter 16), no systematic attempt has been made to understand the phenomena involved in adsorption from multicomponent solutions. The same is also true of adsorption in which a mixture of two or more solids is used. The present account can therefore be no more than a summary of isolated investigations.

ADSORPTION FROM MULTICOMPONENT SOLUTIONS

Most of these investigations have been carried out with ternary mixtures, which have been considered in two groups: two "solutes" in a common "solvent", and one "solute" in a "mixed solvent".

Two Solutes in One Solvent

In general, the presence of a second solute reduces the adsorption of the first. This is readily understood if adsorption is confined to a monolayer and is considered for dilute solutions only, from which the solutes are adsorbed with marked preference over the solvent. Even in multilayer adsorption, the same general result would be expected, as the effect in the first layer would probably play a major part in determining the total adsorption. In such experiments the adsorption of the solvent has generally been neglected. While this is unsatisfactory in a full account of adsorption from ternary mixtures, the results are qualitatively significant, particularly for the more classical cases of adsorption from aqueous solutions by charcoal.

The adsorption of acetone by charcoal from dilute aqueous solution is reduced in the presence of acetic acid,[1] i.e. both acetic acid and acetone are strongly adsorbed relative to water, but with comparable strength relative to each other. Similarly, the adsorption of acetic acid from aqueous solution is reduced in the presence of aliphatic alcohols.

From dilute aqueous solutions, oxalic acid and succinic acid are adsorbed by blood charcoal according to the Freundlich equation. The measured adsorption of each from mixed solutions is considerably less than would occur from a single solution at the same equilibrium concentration. The isotherm for adsorption from mixed solutions still obeys a Freundlich equation, but with different constants.[2]

The proportions in which monocarboxylic acids (aliphatic and aromatic) are adsorbed from aqueous solution by charcoal are not related to the extents of adsorption of the separate solutes.[3] The solutes are thus not "isotatic", as are some substances at the liquid–vapour interface (Chapter 11). In general, the aromatic acids (which were also the less soluble solutes) were more strongly adsorbed than the aliphatic acids. The limiting adsorption from such mixtures corresponded to the coverage of an approximately equal area for all of the mixtures examined.

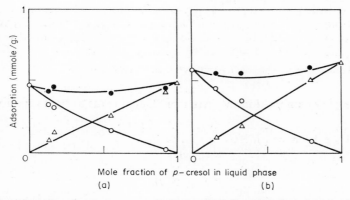

FIG. 9.1. Adsorption from aqueous solutions containing phenol and p-cresol on flue dust[4] at 20°c; $\bigcirc$, phenol; $\triangle$, p-cresol; $\bullet$, total adsorption; points, experimental, curves calculated; total concentration: (a) 3 millimoles/litre, (b) 7 millimoles/litre. (Reproduced from the *Collection of Czechoslovak Chemical Communications*.)

If the solutes are very similar (which includes having similar adsorption isotherms when adsorbed separately), they replace one another in the adsorbed phase almost in proportion to their relative concentrations in the liquid phase. An example is the adsorption on flue dust from aqueous solution of phenol and p-cresol.[4] For a given total concentration, the amount of each component adsorbed by unit weight of solid was almost proportional to the relative mole fraction of that component in the solution (Fig. 9.1), the total adsorption varying very little with composition for a given total concentration.

Thus

$$n_1 = n_{10} x_1 \tag{9.1}$$

and

$$n_2 = \frac{n_{20} x_2}{1 + 0.751 x_1},$$ (9.2)

where n_1 and n_2 are the number of moles of components 1 (phenol) and 2 (p-cresol) adsorbed by unit weight of solid from the mixed solutions, n_{10} and n_{20} are the corresponding amounts adsorbed from separate solutions of the same total concentration, and x_1 and x_2 are the mole fractions of the two components in the total solute, defined by

$$x_1 = \frac{c_1}{c_1 + c_2} \quad \text{and} \quad x_2 = \frac{c_2}{c_1 + c_2},$$

where c_1 and c_2 are the molar concentrations of the respective components in the solution.

For slightly less similar solutes, one component may be preferentially adsorbed with respect to the other. Thus 3,4-dimethyl phenol is adsorbed more strongly than phenol,[5] perhaps because of its lower solubility in water. In this case adsorption of the individual components 1(3,4-dimethyl phenol) and 2 (phenol) could be represented by the equation:

$$n_1 = n_{10} - 0.269 x_1 x_2$$ (9.3)
$$n_2 = n_{20} - 1.166 x_1 x_2.$$ (9.4)

At the higher compositions a phenomenon analogous to the formation of an azeotrope is observed. Erdös and Jäger refer to this as an atropic mixture.

In extreme cases, the adsorption of one component is not affected at all by the presence of a second solute. The adsorption of stearic acid by titania from petroleum ether is unaffected by the presence of considerable excess of short-chain solutes (Fig. 9.2),[6] a phenomenon which appears to be general in long-chain compounds containing a polar group.

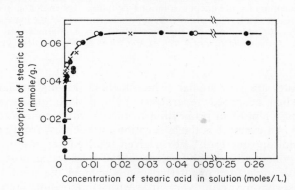

FIG. 9.2. Adsorption on rutile of stearic acid at 25°C. ●, from solution in petroleum ether; ✕, from petroleum ether containing ethanol or phenol; ○, from petroleum ether containing cyclohexylamine.[6] (Reproduced from *Kolloid Zeitschrift* with permission from Dr. Dietrich Steinkopff Verlag, Darmstadt.)

At the other extreme stand pairs of solutes which can interact strongly. Aromatic amines (aniline, o-, m-, and p-toluidine) and aliphatic acids (acetic, propionic, and butyric acids) are adsorbed on charcoal from the mixed aqueous solutions as the free substances, not in combined form.[7] In the presence of excess of acid, the adsorption of the amines is reduced at a given concentration. The extent of reduction depends on the concentration of acid, apparently according to its effect in increasing the solubility of the amine. The effect on solubility is more than offset, as between acetic and butyric acids, by the increased competition of the latter for the surface of the charcoal.

One Solute in a Mixed Solvent

The adsorption of a given solute varies with the composition of the solvent,[8, 9] but the nature of the dependence is not yet clear.

Quaternary Systems

Quaternary mixtures have been considered briefly, e.g. two "solutes" in a mixed solvent. As between the two solutes, preferential adsorption takes place in favour of the component which is the more strongly adsorbed from solutions of the separate solutes in the mixed solvent.[10] As for adsorption from single-component solvents, the more strongly adsorbed solute is usually the less soluble.[11]

ADSORPTION BY MIXTURES OF ADSORBENTS

Adsorption by a mixture of two adsorbents would be expected to be the sum of the adsorption on the two adsorbents separately. Because many adsorbents are heterogeneous (i.e. behave as a mixture of two adsorbents) it is relevant that this situation has been investigated experimentally.

A preliminary confirmation was obtained as a result of some rather simple experiments,[12] though exact results were not obtained at low concentrations because the shape of the isotherm was not adequately taken into account.

REFERENCES

1. L. Michaelis and P. Rona, *Biochem. Z.*, 1908, **15**, 196.
2. H. Freundlich and M. Masius, "Gedenbock aangetoden aan J. M. van Bemmelen", 1910, p. 88.
3. C. Ockrent, *J. Chem. Soc.*, 1932, 613.
4. L. Jäger and E. Erdös, *Coll. Czech. Chem. Comm.*, 1959, **24**, 3019.
5. E. Erdös and L. Jäger, *Coll. Czech. Chem. Comm.*, 1959, **24**, 2851.
6. L. Dintenfass, *Kolloid Z.*, 1958, **161**, 70.
7. E. Angelescu and D. Cismaru, *Bul. Soc. chim. Romania*, 1935, **17**, 229.
8. N. Schilow and S. Pewsner, *Z. phys. Chem.*, 1925, **118**, 361.
9. J. Trividic, *Compt. rend.*, 1928, **186**, 865, 1358.
10. N. F. Ermolenko and S. A. Levina, *Acta Physicochim. U.R.S.S.*, 1939, **10**, 451.
11. N. Ermolenko and M. Bokhvala, *Acta Physicochim. U.R.S.S.*, 1940, **13**, 839.
12. H. Lachs, *Z. phys. Chem.*, 1916, **91**, 155.

CHAPTER 10

Factors Influencing Competitive Adsorption at the Liquid–Solid Interface

INTRODUCTION

In adsorption from binary mixtures, preferential or "selective" adsorption[1] occurs, but to emphasize that both components are adsorbed, it may be preferable to refer to competitive adsorption. From the available evidence it can be seen qualitatively that several factors influence the nature of this competitive adsorption. Clearly the interaction of each component of the solution with the solid surface is important, and to these may have to be added the interaction between the two components of the solution. Other factors, such as the porosity and heterogeneity of the solid, are also significant.

Some of the experimental evidence relates to adsorption at one concentration only. This is unlikely to lead to sound conclusions and may even prove to be misleading. Generalizations are not likely to be satisfactory unless the whole isotherm has been studied.

Although adsorption from solution is usually thought of in terms of physical adsorption, a number of examples of chemisorption (at room

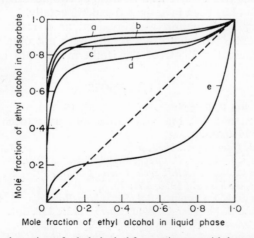

FIG. 10.1. Physical adsorption of ethyl alcohol from mixtures with benzene on (a) gibbsite, (b) silica gel, (c) boehmite, (d) γ-alumina, (e) charcoal.[6] (Reproduced with permission from the "Proceedings of the Second International Congress of Surface Activity", Butterworths.)

temperature) are being found. In binary systems which have been examined so far, only one component has been chemisorbed, and the process has dominated the selective adsorption, even though it has been accompanied by physical adsorption onto the chemisorbed layer. The hydrogen-bonding of one component has an effect intermediate between those of chemisorption and physical adsorption.

Selective adsorption can be expressed in several ways. For comparison of a variety of systems, one of the most convenient is to express the mole fraction of a given component in the adsorbate as a function of the mole fraction of the same component in the liquid phase. An example is given in Fig. 10.1. For selective adsorption to be expressed in this form it is, of course, first necessary to evaluate the individual isotherms.

INTERACTION BETWEEN SOLID AND ADSORBATE

The existence of competition in physical adsorption between the components of a solution depends most frequently on the difference in the strengths of interaction between adsorbent and the adsorbates. (Molecular sieve effects are an example in which this is not so.) It seems likely that quite small differences in adsorptive forces may be responsible for preferential adsorption which can readily be detected experimentally. It is not surprising, therefore, that theoretical treatment of these phenomena is not yet advanced.

In general, the observed phenomena can be understood qualitatively in terms of the varying degrees of polarity shown by the surface and the adsorbates. This can now be taken to a rather greater degree of detail than the simple generalization that polar compounds are more strongly adsorbed than are non-polar compounds by polar solids and that the reverse tends to be true for non-polar solids.

The chemical nature of the surface is of profound importance. Its force field is such that preferential adsorption by a solid is appreciably greater than that which occurs at other interfaces, but the extent, and even the sign, of the selectivity varies considerably. This is most clearly evident when adsorption takes place from a mixture of a polar and a non-polar (or relatively non-polar) liquid. Silica gel adsorbs alcohols in preference to iso-octane[2] and in preference to benzene,[3-6] whereas charcoals usually adsorb benzene in preference to alcohols (Fig. 10.1).[6] These preferences may be ascribed to the highly polar nature of the silica gel surface and the relatively non-polar nature of the charcoal surface. Other gels (alumina, titania gel, etc.) resemble silica gel in this respect.[7, 8] The single polar centre in the molecule of piperidine is sufficient for it to be adsorbed very strongly by these gels in preference to cyclohexane; the two molecules are closely similar except in respect of the polar centre.[9]

On such adsorbents, hydrogen-bonding may be responsible for the preferential adsorption of one component. The relative affinities of silica gel for a series of nitro and nitroso derivatives of diphenylamine and N-ethyl-aniline dissolved in simple solvents depends on the strengths of hydrogen bond which can be formed between adsorbent and adsorbate.[10]

The strong preferential adsorption of phenol from cyclohexane by charcoal is probably largely due to hydrogen-bonding, as the adsorption is drastically reduced by substitution of t-butyl groups in the two ortho positions. The effect of these substituent groups is mainly steric (which alone points to the importance of the hydroxyl group as a centre of adsorption), but indirectly they also weaken the hydrogen bond. Methyl groups produce a much smaller reduction in adsorption.[11]

Charcoals are generally of less definite composition, especially with respect to surface groups,[12,13] than oxide or hydroxide gels. Moreover, the surface can readily be altered by treatment with oxidizing or reducing gases. If a charcoal is very strongly oxidized, its preference for ethyl alcohol relative to benzene increases with oxidation until it approaches that shown by silica gel.[14] Conversely, preferential adsorption of alcohol can, to some extent, be reduced by removal of the oxygen groups from the surface.[15]

The effect of complete removal of oxygen complexes from a carbon surface is shown by comparing adsorption from alcohol–benzene mixtures by a carbon black, Spheron 6, and by Graphon, the same material heated to 2700°C. Graphon shows complete preferential adsorption of benzene, whereas Spheron 6 (which is known to have a considerable concentration of oxygen on the surface[16]) shows preferential adsorption of alcohols over a considerable range of concentration.[17]

It has similarly been shown that the degree of preferential adsorption of octadecanol from benzene increases with the oxygen content of carbon black. Most significantly, the preferential adsorption is immeasurably small if the carbon is heated to remove all traces of surface oxygen.[18] In this case the competition for adsorption sites is virtually between a paraffin chain and an aromatic molecule, and experiments (e.g. on mixtures of cyclohexane and benzene) have shown the aromatic material to be preferentially adsorbed by carbon black.[17]

Among mixtures of polar and relatively non-polar liquids, the system methyl acetate–benzene (Fig. 10.2) presents a less extreme case than an alcohol–benzene mixture. Methyl acetate is preferentially adsorbed by silica gel and by alumina (boehmite), and this can be attributed to specific interaction of the polar centre of the ester molecule with the polar groups which cover the surface of the solid. On charcoal, however, methyl acetate is preferentially adsorbed only at low mole fractions, and over most of the concentration range benzene is preferentially adsorbed. This corresponds to a relatively low concentration of polar groups on the surface of charcoal.[6] Preferential adsorption of methyl acetate by boehmite is almost the same from mixtures with ethylene dichloride as from mixtures with benzene. This implies that the forces responsible for adsorption of these two substances are similar in strength and also smaller than those responsible for the adsorption of methyl acetate.

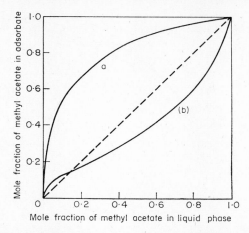

FIG. 10.2. Adsorption of methyl acetate from mixtures with benzene on (a) silica gel or boehmite, (b) charcoal.[6] (Reproduced with permission from the "Proceedings of the Second International Congress of Surface Activity", Butterworths.)

Substances with substituents in the aromatic ring (mesitylene, o- and p-xylene, t-butylbenzene, and isopropylbenzene) are adsorbed much less strongly than benzene at low concentration.[11] The substituent groups have the effect of increasing the distance between the plane of the aromatic ring and the solid surface (if the plane is parallel to the surface) or between the planes of adjacent rings (if the plane is perpendicular to the surface). In either case a reduction in adsorption would occur.

In the absence of specific polar groups, the π-electrons of aromatic systems ensure that aromatic compounds are adsorbed preferentially to corresponding aliphatic compounds by polar solids or any others capable of specific interaction with π-electron systems. A comparison of benzene and cyclohexane (Fig. 10.3) is appropriate because the two molecules are as similar in shape and size as is possible for representatives of the two types of compound.

The preferential adsorption of toluene from n-heptane which occurs on a highly oxidized carbon black is markedly reduced[19] when the oxygen complexes are removed by heating in hydrogen to 1700°C. It would be interesting to examine adsorption from the system benzene–cyclohexane at a non-polar, non-aromatic surface, particularly as cyclohexane is preferentially adsorbed at the liquid–vapour interface (Chapter 11).

An interesting comparison of "chemical" with "geometrical" or "steric" effects is shown in sorption from mixtures of benzene and n-hexane by molecular sieves. The openings of the pores in the Linde molecular sieve 5A are too small to admit molecules of benzene, but molecules of n-hexane are admitted; n-hexane is therefore preferentially sorbed at all concentrations

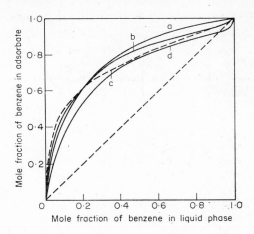

FIG. 10.3. Adsorption of benzene from mixtures with cyclohexane on: (a) Spheron 6, (b) boehmite, (c) charcoal, (d) graphite.[6] (Reproduced with permission from the "Proceedings of the Second International Congress of Surface Activity", Butterworths.)

(see Chapter 4). The sieves 10 X and 13 X, however, have wider openings which admit both molecules. Sorption is thus competitive on a normal basis. The interaction between the π-electron system of the benzene and the ionic lattice of the zeolite is so strong that the n-hexane is completely excluded over virtually the whole range of concentration.[20]

Preferential adsorption is small from the system ethylene dichloride–benzene, which is approximately ideal in the liquid phase.[21] The oxide gels adsorb ethylene dichloride preferentially. This can be attributed to polarization of the chlorine atoms, in which respect it is probably significant that alumina, with a greater ionic character than silica gel, shows a stronger preference for ethylene dichloride. The preferential adsorption shown by charcoal reverses about half-way across the concentration range. This is probably mainly a reflection of the heterogeneity of the charcoal surface.

Importance of Solid–Adsorbate Interactions

The importance of the solid–adsorbate interactions can be seen when adsorption at the liquid–solid interface is compared with adsorption at the liquid–vapour interface for the same liquid system. In the latter case, preferential adsorption is generally governed by the relative volatility of the two components. The sign of preferential adsorption is frequently reversed when the same liquid is in contact with a solid. This is discussed in Chapters 11 and 14. Again, relatively small differences in the energy terms may be sufficient to reverse the selectivity.

Special Features in the Nature of the Adsorbent

Competitive adsorption may be affected by other characteristics of the adsorbent. Apart from its purity (which has been referred to in Chapter 2), the important features are its degree of porosity and the extent of heterogeneity of the surface. The chemical nature of the surface has been considered in the preceding section.

(a) Porosity

Solid adsorbents range from those which are completely non-porous, through those with macro, intermediate, and micro pores to molecular sieves in which the entry to the pores is so small that the larger component is completely excluded though the smaller is admitted. The last category is a limiting case, but as most porous solids have a spread of pore sizes, the possibility that they can show a partial molecular sieve effect should not be overlooked if the components of the liquid mixture are not of the same molecular size. This effect would add a degree of preferential adsorption of the component of smaller size, irrespective of the competitive adsorption due to other factors. The result, as seen in the composite or individual isotherms, would be similar to that due to chemical heterogeneity of the surface.

The general effect of porosity is shown by those types of adsorbent for which a series of related solids can be produced with varying degrees of porosity, e.g. charcoals by varying the degree of activation,[12] silica gels by varying the conditions of forming or calcining the gel.[22] In these cases the increase in porosity is normally accompanied by an increase in specific surface area and thus by an increase in the magnitude of the selective adsorption, e.g. adsorption of fatty acids from water on a series of steam-activated carbon blacks[23] and on steam-activated charcoals.[24]

For carbons which are heated to varying temperatures after activation, adsorptive capacity tends to fall off as the final temperature of preparation[25] exceeds 800°C. This is due to a closure of the pore structure which is general for carbons at temperatures in the range 700°C–1500°C. In general, the effect is one of reduction in adsorptive capacity, but, at intermediate stages, molecular sieve effects may also be observed.

In the practical use of solid adsorbents, porous solids are often important because they have a high adsorptive capacity per unit weight or unit volume. Non-porous solids, even if they have the same inherent selectivity, are of less use because their adsorptive capacity is small unless the solid is finely divided, and in this case it is more difficult to handle.

More specifically, the question arises as to whether adsorption by porous substances differs in any essential way from adsorption by non-porous substances. This can be considered in two ways. One issue is whether adsorption occurs by a pore-filling or an area-filling mechanism. In part this has been dealt with above (Chapter 4). A further example shows the different ways in which experimental results can be interpreted.

Kiselev and Shcherbakova found that the limiting values for adsorption of the lower fatty acids and alcohols from aqueous solution by charcoal decreased with increasing chain-length when expressed in millimoles per gram.[26] (The alcohols are less strongly adsorbed at low relative concentrations but reach the same limiting adsorption.) A similar difference was observed in adsorption on silica gel from solutions of acetic acid and stearic acid, respectively, in carbon tetrachloride.[27] This was assessed against the observation that at the free surfaces of liquids and non-porous solids, the limiting molar adsorption of such compounds is independent of chain-length. For the porous solids, however, the limiting molar *volume* adsorbed was approximately constant for all substances examined, and it was concluded that this volume represented the pore volume of the adsorbent (0·28 c.c./g. in the case of charcoal).

It has subsequently been shown (Chapter 7) that limiting molar adsorption in a homologous series is not always independent of chain-length even on non-porous solids, but depends on the nature of the surface. Thus long-chain alcohols and fatty acids may be adsorbed by polar solids with the chain perpendicular to the surface. The limiting number of moles adsorbed per gram of solid is then independent of chain-length, but the limiting volume of the adsorbate increases with chain-length. On the less polar solids, these solutes may be adsorbed with the chain parallel to the surface. When a complete monolayer is formed, the area covered is the same for all members of the series (but the number of moles adsorbed per gram varies), and the limiting volume of adsorbate remains approximately constant.

A constant volume of adsorbate in adsorption of members of a homologous series on a porous solid need not, therefore, imply that a pore-filling mechanism is operative. The result is compatible with monolayer adsorption with the chain parallel to the surface. For some solids (e.g. a silica gel with wide pores) the volume of a monolayer would be much smaller than the total pore volume. For others (e.g. many activated charcoals with narrow pores), the difference might be very small as most of the surface area is located in pores too narrow to adsorb more than one molecular layer.

Adsorption on the charcoal used by Kiselev[25, 26] was probably confined to a monolayer because the amount of methyl alcohol adsorbed from the vapour phase just *before* the onset of capillary condensation was also 0·28 c.c./g. expressed as a liquid volume. Furthermore, the limiting adsorption of other solutes varied: 0·04 c.c./g. (Congo red), 0·12 (succinic acid), 0·17 (dimethyl ethyl carbinol), 0·19 (salicylic acid), 0·20 (methylene blue), 0·22 (benzoic acid), 0·23 (cyclohexanol), 0·26 (phenol). In some cases the low value could be attributed to incomplete penetration by a large molecule into all of the pore system. This is unlikely to be the case with all of the substances listed; other possibilities are that the adsorbed molecules are of such a shape as to leave extensive gaps in the monolayer even at closest packing, and (possibly in the case of succinic acid) the existence of solvent molecules in the adsorbed layer even at limiting adsorption. The low value for Congo red probably results from the inability of the semi-colloidal material to enter the pores of the charcoal.

Care is still needed in interpreting results for porous charcoals. Studies of adsorption of vapours suggest that the pores may consist of relatively large cavities to which access is obtained through quite small openings or "windows".[28] The "windows" may be slits opening into spaces between

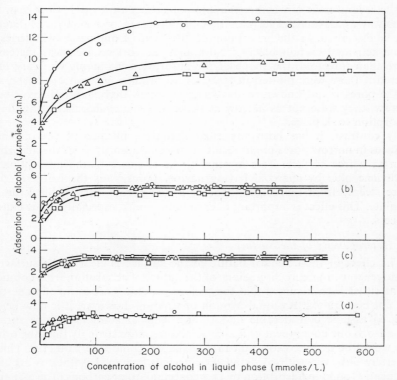

FIG. 10.4. Adsorption of aliphatic alcohols from carbon tetrachloride by silica gels of different degrees of hydration[30] (○, △, and □): (a) methyl alcohol, (b) propyl alcohol, (c) hexyl alcohol, (d) octyl alcohol. (Reproduced from *Doklady Akad. Nauk. S.S.S.R.*)

graphite layers rather than circular or square openings leading to spherical cavities.[29] The relevance of this model to adsorption from solution has not yet been fully assessed.

The adsorption of aliphatic alcohols from carbon tetrachloride by non-porous silica gel shows that two different factors can be present in adsorption of a homologous series. For the lower members, the number of moles of methyl alcohol adsorbed by unit area of surface is much higher for a highly hydrated silica than for a less highly hydrated silica (Fig. 10.4).[30] For n-octanol, the difference can only be observed at very low concentrations. Moreover, the limiting adsorption decreases with increasing chain-length of the alcohols. This suggests that the adsorption of the methyl alcohol molecule depends primarily on interaction of its hydroxyl groups with those of the silica surface, whereas the octanol molecules are adsorbed primarily because the hydrocarbon chain is held by the surface (with the major axis parallel to the surface).

The second issue is whether the internal surface of a solid behaves in the same way as an external surface. In some instances, no significant difference can be discerned. In the adsorption of fatty acids from water by a series of steam-activated carbon blacks,[23] although the surface area was increased more than threefold, the shape of the adsorption isotherms was very little altered if these were plotted in terms of adsorption per unit area. The size of the pores is not known in this case; although they may well have been narrow, they were not as narrow as those of some charcoals, as multilayer adsorption took place.

By contrast, it has been suggested that the influence of two adjacent surfaces in narrow pores might result in stronger adsorption of the adsorbate than at a free surface. This appears to be the case in the adsorption of methylene blue by charcoal having narrow pores,[31] as the isotherm rises more steeply from the origin than for a non-porous carbon black or for graphite. This effect is important in chromatography, especially in the elution process.

(b) Heterogeneity of the Surface

The effect of the chemical nature of the surface is clear if the surface is homogeneous. In many discussions of the nature of adsorption from solution, it has either been assumed that the surface is homogeneous, or the question has not been raised. As in adsorption of vapours, however, it is now beginning to assume considerable importance. Thus if a homogeneous surface can adsorb one component of a binary mixture more strongly than the other, this would be expected to result in a U-shaped composite isotherm. This is generally observed with oxide adsorbents, but S-shaped isotherms are very frequently found in adsorption by charcoals and other carbons. The S-shaped isotherm can be attributed to more than one factor (see p. 177). It would be reasonable to suppose, however, that it would arise if two types of chemical group were present on the surface of the adsorbent, having different affinities for the two components of the liquid. This is shown schematically in Fig. 4.10.

Heterogeneity in oxide surfaces is not always important. It generally consists in the presence of both oxide and hydroxide groups (as, for example, in boehmite) and as both are highly polar, their different discrimination between pairs of adsorbates is not always easy to detect. It is thought to be important, however, in the adsorption of fatty acids by silica gel, as the occurrence of hydrogen-bonding is postulated:

$$\ce{-Si} \begin{array}{c} \diagup O \cdots H \!-\!\!-\! O \diagdown \\ \diagdown O \!-\!\!-\! H \cdots O \diagup \end{array} \ce{C-CH3,}$$

analogous to the formation of a dimeric molecule of fatty acid.[32] The double hydrogen-bond could not be formed on a surface consisting of oxide groups only. A similar effect is important in determining the spacing between stearic acid molecules adsorbed from solution in organic solvents on various kinds of alumina.[33]

It is now recognized that charcoal is a very varied substance (or that the term includes a family of substances[11,12]) and that the same applies to carbon black, for which a more definite classification is available.[34,35] In both cases, the surface is normally heterogeneous to an extent which has not been sufficiently taken into account. It is useful to consider the surface not as elementary carbon, but as a polynuclear hydrocarbon,[36] interspersed with a variety of oxygen complexes,[12,13] and in some cases with groups containing other elements (e.g. nitrogen and sulphur). The extent of heterogeneity varies according to the method of manufacture of the carbon and its subsequent thermal history.

In the examples quoted above, the selective adsorption of methyl acetate from benzene at low concentrations can be attributed to the strong interaction of the ester group with some of the polar oxygen complexes. When these are saturated, however, adsorption on the remainder of the surface is governed by the π-electron interaction between the aromatic solid and the aromatic benzene molecules, which is stronger than that between the solid and the aliphatic ester. In the same way the oxygen complexes probably have a polarizing influence (similar to, but smaller than that of the oxygen groups of silica gel) on the chlorine atoms of ethylene dichloride, thus leading to preferential adsorption of ethylene dichloride at low concentrations, whereas benzene is again preferentially adsorbed at higher concentrations.

A more detailed examination of adsorption by charcoal from mixtures of the lower aliphatic alcohols with benzene [37] suggests that the small degree of preferential adsorption of the alcohols is due to their adsorption (possibly by hydrogen-bonding) by oxygen complexes. The individual isotherms (Fig. 10.5) strongly resemble those shown in Fig. 4.10. Taken together, they show that the number of moles adsorbed in this way decreases as the homologous series is ascended, possibly reflecting the effect of increasing molecular size. Over the rest of the surface, presumed to be of hydrocarbon character, the benzene is preferentially adsorbed in all cases.

The same phenomenon is shown in adsorption from the same systems by carbon blacks.[17] On Spheron 6, which has a heterogeneous surface, n-butyl alcohol is preferentially adsorbed at low concentrations, but not at high concentrations. On Graphon, which has no appreciable heterogeneity, benzene is preferentially adsorbed at all concentrations (Fig. 10.6).

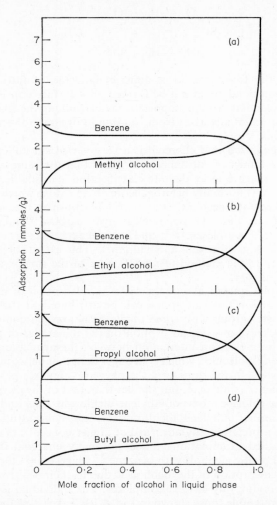

FIG. 10.5. Adsorption (individual isotherms) from mixtures of alcohols and benzene on charcoal[37]: (a) MeOH—C_6H_6, (b) EtOH—C_6H_6, (c) PrOH—C_6H_6, (d) n-BuOH—C_6H_6. (Reproduced with the permission of the American Chemical Society from the *Journal of Physical Chemistry*.)

It may become necessary to consider a special kind of heterogeneity in ionic crystals—the presentation of different planes at the adsorbing surface. There is evidence that urea is not adsorbed in the same way from aqueous solution by the 100 and 111 planes of sodium chloride.[38]

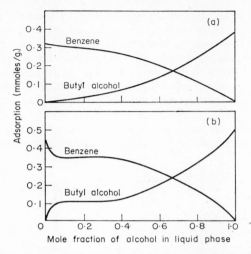

FIG. 10.6. Adsorption (individual isotherms) from mixtures of n-butyl alcohol and ben-zene[17] by: (a) Graphon, and (b) Spheron 6. (Reproduced with the permission of Pergamon Press from "Proceedings of the Fourth Conference on Carbon".)

INTERACTION IN THE LIQUID STATE

This factor was ignored in early work, but is now increasingly seen to be important. In the simplest sense, it is responsible for the difference in behaviour between completely and partially miscible systems. In this limiting case of a solute sparingly soluble in a solvent, it is sometimes useful to think of adsorption as being a distribution of the solute between the liquid phase and the adsorbed phase.

Completely Miscible Liquids

For completely miscible systems, the gross effects are readily detectable in suitably chosen experiments. Thus on the homogeneous surface of Graphon, benzene is preferentially adsorbed from both methyl alcohol and n-butyl alcohol. This seems to be due in part to the nature of the interaction at the solid surface (see p. 167). The marked difference in degree of preferential adsorption as between the two systems, however, suggests that another factor is important.[17] The molecule of methyl alcohol is so small that the polar hydroxyl group forms a large fraction of the whole. This group cannot interact strongly with the solid surface, but can do so, by hydrogen-bond formation, with the hydroxyl groups of other molecules and thus has a tendency to remain in the liquid phase (where the other hydroxylic molecules are to be found) instead of being adsorbed by the surface. This interaction between the methyl alcohol molecules tends to exclude the benzene molecules which, being in any case more strongly held by the surface, readily move to the adsorbed layer. The liquid system comes close to phase separation at room temperature, as is shown by the partial pressure curves (Fig. 10.7), and the multilayer formation observed at the solid surface therefore appears to be incipient phase separation. In butyl alcohol the hydroxyl group forms a smaller part of the molecule, and the longer hydrocarbon chain makes for readier miscibility with benzene. There is consequent readier retention of benzene in the liquid phase and a reduced preferential adsorption of benzene. Further, the increasing number of methylene groups in the butyl alcohol molecule results in its being more strongly adsorbed than methyl alcohol at the solid surface, and therefore in its competing more effectively with benzene.

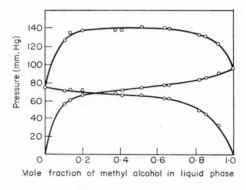

FIG. 10.7. Partial pressure curves and total pressure curve[17] for mixtures of methyl alcohol and benzene at 20°C. (Reproduced with the permission of Pergamon Press from "Proceedings of the Fourth Conference on Carbon".)

If hydrogen-bonding occurs between two components which are completely miscible in the liquid phase, it may have little effect on the competitive nature of adsorption. Thus competitive adsorption from mixtures of pyridine and ethyl alcohol shows only a slight departure from the behaviour which would be expected in the absence of hydrogen-bonding.[39] Nor are any special effects noticed when there is weak hydrogen-bonding (as between chloroform and acetone) in the liquid phase, even though it may also occur in the adsorbed phase.[40]

Adsorption by charcoal from the pyridine–water system has been described in Chapter 4. This shows that when one component is strongly adsorbed (in this case, pyridine), and the second is weakly adsorbed, enhanced adsorption of the second component may result from its attachment, by hydrogen-bonding, to the component which is strongly held by the surface.

The effect of a moderate degree of interaction in the liquid phase has been considered by Jones and Mill.[41] They regard the vapour phase as the ideal phase of the binary mixture and plot the composition of both condensed phases (the liquid phase and the adsorbate) against the composition of the vapour (Fig. 10.8). For a series of mixtures adsorbed on silica gel, the two condensed systems are closer to one another in composition than either is to the vapour. The extent to which this is so may be somewhat exaggerated, as no analysis was made to distinguish between primary adsorption and filling of capillaries; to the extent to which the latter occurs, the composition calculated for the adsorbate is brought closer to that of the bulk liquid.

The liquid systems show considerable deviations from Raoult's Law. In two cases, the systems form an azeotropic mixture. Adsorption from these systems gives individual isotherms which have a point of inflexion if plotted against the composition of the liquid. When they are plotted against vapour composition, however, there is no inflexion. Combination of these individual

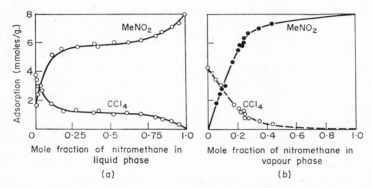

FIG. 10.8. Adsorption (individual isotherms) on silica gel from mixtures of nitromethane and carbon tetra-chloride plotted as a function of the composition of: (a) the bulk liquid, and (b) the vapour phase mixture.[41] (Reproduced with permission from the *Journal of the Chemical Society.*)

isotherms would then give a U-shaped instead of an S-shaped composite isotherm. In such cases the shape of the adsorption isotherm is determined, at least in part, by the interaction of the two components in the liquid phase.

The individual isotherms for adsorption from water–ethyl alcohol mixtures showed inflexions, even when plotted against vapour composition. This suggests exceptional interaction between the adsorbate and the surface; for both components slow chemisorption on silica gel is possible,[8, 42] and this system may therefore not be strictly comparable with the others which Jones and Mill examined.

Other cases in which the composite isotherm remains S-shaped, even when plotted against composition of the vapour, include adsorption by charcoal from mixtures of the lower alcohols and benzene, but this probably results from chemical heterogeneity of the surface, as described above (p. 173).

Komarov and Ermolenko have attempted to summarize the effect of deviations from Raoult's Law as follows: if the solution shows negative deviation from ideality, the component present in excess is selectively adsorbed; if the solution shows positive deviation, the component present in lower concentration is selectively adsorbed.[43] Although this generalization can be drawn from the results quoted by Komarov and Ermolenko, it does not seem to apply to all other results. This has also been expressed for regular solutions in terms of the tendency for the component present in the smaller amount to "escape" from the solution (and thus be preferentially adsorbed) when mixing of the two components is endothermic. If the two components interact equally strongly with the surface, an S-shaped composite isotherm results, as the component present in low mole fraction at each end of the concentration range "escapes" from the solution.[44]

From thermodynamic and kinetic considerations, Elton has concluded that cases of complete preferential adsorption (U-shaped composite isotherms) will rarely be found.[45] It is now clear, however, that this does often occur, and it would be useful to re-consider the thermodynamic implications. In the most recent theoretical work it is apparent that, for mixtures close to ideality, the nature of the interaction in the liquid phase may be the dominant factor influencing the adsorption isotherm. In the limit, in which the two components are adsorbed equally strongly, it can be visualized as the only factor. This is discussed further in Chapter 14, in which the conditions in which liquid-phase interactions can give rise to an S-shaped composite isotherm are considered further.

Systems of Limited Concentration Range

The important effects have been discussed in Chapters 5 and 7. Briefly, a complete monolayer of one component is usually formed quite readily in adsorption from partially miscible liquids. At higher relative concentrations, multilayers of this component are often formed, and this process can be regarded as incipient phase separation. From solutions of solids, a complete monolayer may not be formed even at the highest available concentration if the solid is present in a good solvent which is also strongly adsorbed by the surface. It may, however, form a complete monolayer, and in a few cases, exceed this at high relative concentrations.

There are indications that some completed monolayers formed on a homogeneous surface (e.g. iodine or stearic acid on Graphon) have an ordered structure very similar to that which exists in a single layer of the corresponding crystal. The ordered packing may, in part, account for the greater readiness to form a complete monolayer on a homogeneous surface than on a heterogeneous surface (e.g. Spheron 6) on which a more disordered packing would almost certainly result from random distribution of the more and the less active sites. If the binding in the adsorbed layer is very strong (e.g. that in hydrated oxalic acid, in which extensive hydrogen-bonding to form a two-dimensional network is postulated), ordered packing appears to occur on both homogeneous and heterogeneous surfaces (Graphon and Spheron 6).

When other effects are not present, a given solute is generally more strongly adsorbed from poor solvents than good solvents. An attempt has been made to relate this to the dipole moment of the solvent,[46] but this approach has received little support. Similarly, for a given solvent, the less soluble solutes are generally more strongly adsorbed than the more soluble. This is most likely to be the explanation of the results for adsorption by charcoal from aqueous solution of a large number of organic acids.[47] Increasing the proportion of polar groups in the compound (and so increasing solubility) decreased adsorption. As the decrease was observed at low concentrations as well as high concentrations, the increasing molecular size due to the addition of side-groups is unlikely to have been responsible for the reduced adsorption. A similar reduction in adsorption of phenols from aqueous solutions on charcoal occurs on increasing the number of (hydrophilic) hydroxyl groups in the molecule.[48] It has also been observed that, on charcoal, the more hydrophilic substances are more strongly adsorbed from hydrophobic than from hydrophilic solvents, and vice versa.[49]

An attempt has been made to relate adsorption to solubility in a quantitative way. Patrick and Eberman proposed the equation[50]:

$$V = k \left(\frac{S\gamma}{S_0}\right)^{1/n}, \tag{10.1}$$

where V is the volume of the liquid solute adsorbed by unit weight of adsorbent, γ is the interfacial tension, S_0 is the solubility, and S the equilibrium concentration of the solute in the bulk liquid, k and n being constants. Adsorption of naphthalene and of acenaphthene on charcoal from a variety of solvents tends to decrease as γ/S_0 decreases, but in an irregular way.[51]

Homologous Series: Traube's Rule

(a) Homologous Series in Adsorption

The regularities which are often found in the simpler physical properties of homologous series also extend to their behaviour at interfaces. The first important generalization was made by Traube on the surface tensions of aqueous solutions of organic solutes. He found that surface activity increased strongly and regularly as any series was ascended.[52] The quantitative formulation of this finding is now known as Traube's Rule, and is discussed further for the liquid–vapour interface in Chapter 11.

Because surface activity is related to adsorption at the interface, Traube's Rule has been expressed in terms of the regular increase in the strength of adsorption of successive members, as measured by the work required to remove one mole from the surface layer to the bulk of the solution.

On this basis, it is interesting to compare the behaviour of homologous series at other interfaces. Unfortunately, few reliable data are available for energies of adsorption at the liquid–solid surface. More numerous observations have been made on the extent of adsorption of a few members of a homologous series by a given solid and from the same solvent. There has been a tendency for any progressive change to be regarded as an example of Traube's Rule (or as an inversion of it) whether or not a quantitative relationship is obeyed. The original concept of the decreasing concentrations required to produce a given degree of surface activity (equated with adsorption) has been supplemented by the concept of increasing extent of adsorption for a given concentration. This latter concept has sometimes been used very loosely, particularly when it is ignored that Traube's Rule should strictly be regarded as a limiting case applied to infinitely dilute solution. It is, nevertheless, interesting to examine the behaviour of homologous series as widely as possible.

(b) Adsorption from Dilute Solutions

Freundlich's early work on adsorption of the lower fatty acids from aqueous solution by charcoal[53] showed (Fig. 10.9) increasing adsorption with increasing chain length. From decimolar solution, the molar ratios of acid adsorbed per gram of charcoal were, acetic : formic, 1·26; propionic :

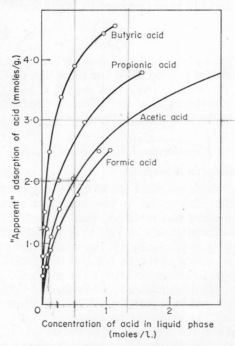

FIG. 10.9. Adsorption by charcoal from dilute aqueous solutions of the lower fatty acids.[53] (Reproduced with permission from "Colloid and Capillary Chemistry", Methuen.)

acetic, 1·55; n-butyric : propionic, 1·56. Similar sets of isotherms have been obtained by many other investigators. Some of the earlier results are quantitatively suspect, because purified charcoals were not always used (cf. Chapter 2). It has also been suggested that ion-exchange could occur with charcoals having a high surface concentration of oxygen complexes,[54] though this effect can be exaggerated; Fig. 10.10 shows results obtained with ash-free charcoal.[55]

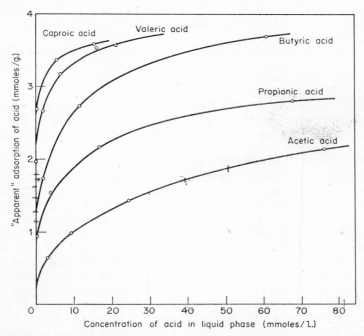

FIG. 10.10. Adsorption of the lower fatty acids from aqueous solution on ash-free sugar charcoal.[55] (Reproduced with permission from *Zeitschrift für Physikalische Chemie.*)

These results can be interpreted along the lines put forward by Langmuir for the liquid–vapour interface of the same solutions (Chapter 11), namely that the molecules of the acid were adsorbed with the major axis parallel to the surface, each CH_2 group contributing an equal amount to the energy of adsorption.

The investigation has been extended to higher members of the series by Claesson,[56] using frontal analysis. The solutions were necessarily dilute, and the isotherms could be fitted by a Langmuir equation, relating adsorption (a^0) to concentration (c), of the form:

$$a^0 = \frac{kc}{1+lc}.$$ (10.2)

The constant k varied regularly, and a linear plot of $\log k$ against n (the number of carbon atoms in the molecule) was obtained for the mono-carboxylic acids ($n = 6$ to 18), the dicarboxylic acids ($n = 7$ to 12), the alcohols ($n = 7$ to 16), and the ethyl esters (ethyl decanoate to ethyl docosanoate). The curves for acids and alcohols were not identical (in contrast to Guastalla's findings for the liquid–vapour interface, Chapter 11), but the differences were small. For a given value of n, acids were adsorbed to decreasing extents in the order: saturated acids (straight-chain acids adsorbed more than branched-chain acids) > *trans*-unsaturated acids > *cis*-unsaturated acids.

Other systems to which Traube's Rule applies include solutions of paraffin wax in benzene, with charcoal as adsorbent,[57] and acidified aqueous solutions of aliphatic aldehydes with charcoal.[58] The results for the last system may be of biological significance.

Aqueous solutions of dicarboxylic acids show anomalous results in adsorption by charcoal;[59] this is paralleled by adsorption at the liquid–vapour interface for the same systems.

(c) Adsorption from Concentrated Solutions

Although Traube's Rule applies strictly to dilute solutions, to which the earlier experiments were mainly confined, it is interesting to examine the behaviour of homologous series in more concentrated solutions. The complete isotherms for adsorption on charcoal from aqueous solutions of the first four fatty acids have the form shown[60,61] in Fig. 10.11. As plotted, the increasing maxima suggest a qualitative extension of Traube's Rule to higher concentrations. This, however, is a fortuitous result of expressing adsorption in terms of weight units ($w_0 \Delta c/m$). The corresponding values of $n_0 \Delta x/m$ (cf. Chapter 3) are (formic) 3·19, (acetic) 3·02, (propionic) 3·18, (butyric) 3·01 millimoles per gram.† The approximate constancy of these figures may be of no significance; they are surface excesses at concentrations too high to be approximately equal to the actual surface concentrations of the acids.

The fatty acid molecules are assumed to be oriented in the same way whether adsorbed from solutions of high or low concentration. In adsorption of the mixed vapours (of fatty acid and water) the fatty acid molecules are adsorbed with the hydrocarbon chain parallel to the surface, and this orientation is assumed to be adopted at the liquid–solid interface, in contrast to the orientation adopted at the liquid–vapour interface.

At high concentrations, the number of molecules of fatty acid adsorbed by unit weight of solid decreases with increasing chain-length, but the area covered increases, and that available to the water molecules decreases. Thus the operation of Traube's Rule at low concentrations has a continuing

† The plot of $n_0 \Delta x/m$ is based on nominal mole fractions. Aqueous solutions of the lower fatty acids are so complex[62] that nominal mole fractions are far from an exact representation of the molecular composition of the solutions.

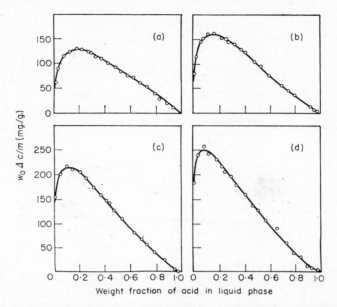

FIG. 10.11. Adsorption (composite isotherms) by coconut shell charcoal from aqueous solutions of: (a) formic acid, (b) acetic acid, (c) propionic acid, (d) butyric acid.[60] (Reproduced with permission from the *Journal of the Chemical Society.*)

effect at the higher concentrations in that the longer the chain-length of the acid, the more effectively it competes with water for the solid surface. This is shown best by plotting the individual isotherms (Fig. 10.12).

This explanation is similar to that given by Langmuir[63] for the adsorption of these acids at the liquid–vapour interface (see Chapter 11). From dilute solutions, the individual isotherms show that the higher acids are more strongly adsorbed, at a given concentration, than the lower acids. This can be attributed to the higher heat of adsorption which is associated with the greater number of methylene groups if the hydrocarbon chain lies along the surface. It would be more difficult to explain these results if the molecule were oriented perpendicular to the surface, when the same group would be presented to the surface irrespective of the chain-length of the acid.

Miculicich has considered the significance of Traube's Rule in terms of the Freundlich equation for adsorption of the solute.[58] He showed that the rule is quantitatively valid at all concentrations only if the exponential term in the Freundlich equation is the same for all members of the series. Otherwise it is only valid over a narrow range of concentration. It would be interesting to extend this approach to cover other equations for the isotherm.

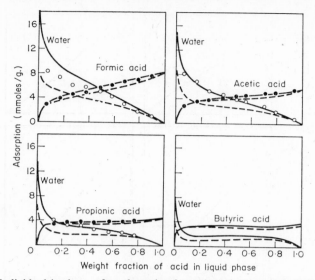

FIG. 10.12. Individual isotherms for adsorption by activated coconut shell charcoal from aqueous solutions of the first four fatty acids[60]: ●, vapour-phase results (acids); ○, vapour-phase results (water); – – – calculated from liquid-phase results on the assumption that water is adsorbed as monomer; —— calculated from liquid phase results on the assumption that water is adsorbed as dimer. (Reproduced with permission from the *Journal of the Chemical Society*.)

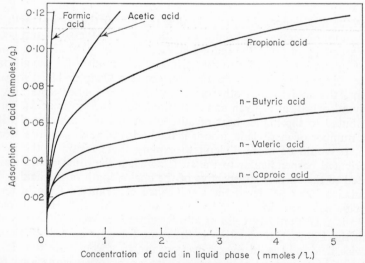

FIG. 10.13. Adsorption of fatty acids from aqueous solution by charcoal prepared under strongly oxidizing conditions—"reversal of Traube's Rule".[64] (Reproduced with permission from *Zeitschrift für Physikalische Chemie*.)

(d) Reversal of Traube's Rule

· Some systems appear to show a reversal of Traube's Rule in the qualitative sense; that is, the extent of adsorption decreases as the homologous series is ascended. An example (Fig. 10.13) is the adsorption of fatty acids by sugar charcoal from aqueous solutions.[64] Nekrassow attributed this reversal to the polar nature of the charcoal, which had been prepared under strongly oxidizing conditions. This would result in increasingly effective competition by water for sites on the surface. Traube's Rule had, in this qualitative sense, originally been applied to the adsorption of a solute from a polar solvent (water) by a relatively non-polar solid (many charcoals). Holmes and McKelvey expected a reversal of Traube's Rule in adsorption from a relatively non-polar solvent by a polar solid.[65] This they found in adsorption of fatty acids from toluene by silica gel; it is even found from 0·01N aqueous solutions.[66] Another example is found in the adsorption of nitro-paraffins from benzene on silica gel.[41]

As most charcoals and gels are porous, the reversal might have been due to exclusion of the larger molecules from very narrow pores[67] (molecular sieve action in present nomenclature). According to Bruns,[68] it is this, rather than polarity of the surface, which accounts for Nekrassow's results. Heymann and Boye established that a reversal of Traube's Rule was observed in adsorption of fatty acids on gold powder which they assumed to be non-porous.[69] Although adsorption was carried out at only one concentration (0·01 M), the reversal was shown for several solvents.

Crisp has pointed out that the supposed reversal of Traube's Rule arises from a confusion between total amount adsorbed and energy of adsorption.[70] On the rather scanty evidence available, he doubts whether in any series the energy of adsorption decreases with increasing chain-length. In this he is following Langmuir's interpretation of Traube's Rule (see Chapter 11), and considering only those cases in which the same area of solid is available to all members of the homologous series.

From this point of view, adsorption of fatty acids from aqueous solution by non-porous carbon blacks is interesting. If the isotherms (Fig. 10.14) are plotted as a function of concentration,[71] Traube's Rule is followed in the qualitative sense. When plotted as a function of activity, they are congruent. Similar congruence was observed in the adsorption of the fatty alcohols from aqueous solutions. Rather less satisfactory congruence, especially at low activities, was found in the adsorption of paraffins from methanol.[72]

Hansen and Craig explained these results by citing Hansen, Fu, and Bartell's comment[73] that the work required to remove a solute from solution as pure solute depended only on the activity of the solute. This would lead to congruence of isotherms for a homologous series if the adsorbent acted on the same functional group. This interpretation would provide a basis both for Traube's Rule and for its reversal. Thus Hansen and Craig predicted a

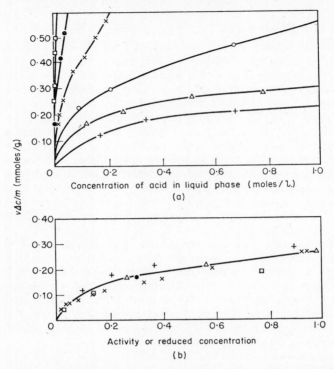

FIG. 10.14. Adsorption by Spheron 6 from aqueous solutions of the lower fatty acids,[71] plotted as a function of (a) concentration, (b) activity: +, acetic acid; △, propionic acid; ○, n-butyric acid; ×, n-valeric acid; ●, n-caproic acid; ☐, heptanoic acid. (Reproduced with the permission of the American Chemical Society from the *Journal of Physical Chemistry*.)

reversal in the adsorption of the lower fatty acids and alcohols from hydrocarbon solvents because the activity coefficients would decrease on ascending the series.

The results obtained by Holmes and McKelvey go some way to establishing this point, although a non-porous adsorbent was not used, and no attempt was made to establish a quantitative relationship for the homologous series. A similar effect was later observed[17] in the adsorption of fatty alcohols from benzene by Spheron 6, but the conclusion is only qualitative because the investigation was not primarily directed to dilute solutions. The normal order appears to apply in adsorption by Graphon. In this case the adsorbent probably does not act on the same functional group. At low concentrations, Spheron 6 probably attracts the hydroxyl group, whereas Graphon has not sufficient polar sites to do so. Hansen and Craig's suggestion does, however, suggest a way to a new interpretation of Traube's Rule.

Against this background, it seems that some of the earlier results can be seen to show mainly the effect of surface structure, though the adsorbents were rarely characterized. Thus with some charcoals the normal behaviour was observed with solutions of fatty acids in water, but a reversal with solutions in carbon disulphide and intermediate behaviour with a number of oxygen-containing solvents such as alcohols and acetone (reversal for the lower members, but normal behaviour for the higher members).[74] This is apparently the result of transition in the solvent from highly polar to relatively non-polar character, with a highly polar charcoal.

ORIENTATION OF THE ADSORBED MOLECULES

The orientations adopted by the molecules of the adsorbate are a consequence of the interaction of the preceding factors. Orientation is, however, of sufficient importance to justify a separate section. If a molecule is highly unsymmetrical, the orientation which it adopts may considerably alter the number of molecules which can occupy unit area of surface, and hence the magnitude of $\frac{n_0 \Delta x}{m}$. For example, the molecule of stearic acid when adsorbed with its hydrocarbon chain along the surface occupies approximately five times the area which it occupies when the chain is perpendicular to the surface. It adopts the latter orientation only on very polar surfaces or when chemisorbed. In such cases a non-polar solvent is almost excluded from the surface. The adsorption of the acid can therefore be detected on solids of much lower specific surface area than is possible when orientation occurs parallel to the surface.

Just as it is impossible to determine directly how much of a given component is present in the adsorbed phase, so its orientation in that phase cannot be determined directly. In a few special cases (for example, long-chain fatty acids) authors have assumed that the orientation adopted at the solid–liquid interface is the same as that adopted in close-packed films on aqueous substrates, and this assumption has sometimes been extended to apply to the lower members of the same homologous series.[75] This now appears to be an unsatisfactory assumption for surfaces which are much less polar than water (Chapter 7).

A better solution is to determine the orientation adopted by a given substance in vapour-phase adsorption on the solid concerned. This cannot be guaranteed to be the orientation adopted in adsorption from solution, but the circumstances are as close as can, so far, be obtained. It is therefore useful to explore how far this assumption leads to reasonable interpretation of liquid-phase data.

The importance of orientation in the interpretation of Traube's Rule has been referred to in the preceding section.

Steric Effects

In the adsorption of polycyclic compounds, steric factors may be important with iso-octane as solvent, and alumina as adsorbent. Adsorption increases with (a) the number of double bonds in the molecule, (b) the extent of conjugation, (c) increasing approach to co-planarity, (d) increasing symmetry number, i.e. the number of equivalent orientations possible for adsorption with the plane of the molecules parallel to the surface.[76]

Another steric factor may be important in the adsorption of substituted phenols by charcoal from solution in cyclohexane. The adsorption of phenols with substituents in the 2,6 positions is much less than that of unsubstituted phenol, suggesting that the steric effect of the substituent groups reduces the possibility of hydrogen-bonding to the surface through the hydrogen group.[11] The effect is not completely established because substituted benzenes are much less strongly adsorbed than is benzene itself. Moreover, as the charcoal was porous, a general molecular sieve effect, as opposed to a specific steric effect, should also be considered.

Stereo-isomers would be expected not to differ greatly in the extent to which they are adsorbed. The differences which exist are none-the-less sufficient to enable separations to be accomplished with chromatographic columns. This applies both to optical and to *cis–trans* isomers. The subject has been reviewed by Zechmeister.[77] An obviously important field is that of the carotenoids.

EFFECTS OF TEMPERATURE AND PRESSURE

Temperature

The detailed effects of temperature have been described in the preceding chapters. They can be summarized as follows.

For adsorption from completely miscible liquids, selectivity generally decreases with rise in temperature. With fall in temperature, multilayer adsorption is likely to occur as the critical solution temperature is approached.

For adsorption from liquids of limited miscibility or from solutions of solids, the general effect of rise in temperature is to reduce selectivity. The rise in temperature usually increases the limit of solubility, sometimes resulting in greater adsorption because higher concentrations become available. In some cases, however, the lowering of solubility with rise in temperature produces opposite effects. Solutions of polymers constitute a special case.

For some systems it has been shown that multilayer adsorption of one component (incipient phase separation) occurs if it is adsorbed above its melting point, but monolayer adsorption occurs (formation of a single pseudo-crystalline layer) if it is adsorbed below its melting point.

Pressure

This factor is generally of little interest. It has, however, been examined by Rosen, who has shown that the adsorption of acetic acid by charcoal from aqueous solution is increased slightly as the external pressure is increased from 1 to 2000 atm.[78] No explanation of this phenomenon has been put forward.

REFERENCES

1. B. S. Rao, *J. Phys. Chem.*, 1932, **36**, 616.
2. F. E. Bartell and F. C. Benner, *J. Phys. Chem.*, 1942, **46**, 847.
3. F. E. Bartell, G. H. Scheffler, and C. K. Sloan, *J. Amer. Chem. Soc.*, 1931, **53**, 2501.
4. D. C. Jones and L. Outridge, *J. Chem. Soc.*, 1930, 1574.
5. F. E. Bartell and G. H. Scheffler, *J. Amer. Chem. Soc.*, 1931, **53**, 2507.
6. J. J. Kipling, "Proceedings of the Second International Congress of Surface Activity", Butterworths, London, 1957, Vol. III, p. 462.
7. J. J. Kipling and D. B. Peakall, *J. Chem. Soc.*, 1956, 4828.
8. J. J. Kipling and D. B. Peakall, *J. Chem. Soc.*, 1957, 4054.
9. J. J. Kipling and D. B. Peakall, *J. Chem. Soc.*, 1958, 184.
10. W. A. Schroeder, *J. Amer. Chem. Soc.*, 1951, **73**, 1122.
11. O. H. Wheeler and E. M. Levy, *Canad. J. Chem.*, 1959, **37**, 1235.
12. J. J. Kipling, *Quart. Rev.*, 1956, **10**, 1.
13. R. N. Smith, *Quart. Rev.*, 1959, **13**, 287.
14. F. S. Bartell and L. E. Lloyd, *J. Amer. Chem. Soc.*, 1938, **60**, 2120.
15. B. W. Puri, S. Kumar, and N. K. Sandle, *Indian J. Chem.*, 1963, **1**, 418.
16. W. R. Smith and W. D. Schaeffer, "Proceedings of the Second Rubber Technology Conference", London, 1948.
17. C. G. Gasser and J. J. Kipling, "Proceedings of the Fourth Conference on Carbon", Pergamon Press, London and New York, 1960, p. 55.
18. J. C. Abram and G. D. Parfitt, "Proceedings of the Fifth Conference on Carbon", Pergamon Press, London and New York, 1962, p. 97.
19. A. V. Kiselev and V. V. Platova, *Zhur. fiz. Khim.*, 1956, **30**, 2610.
20. S. P. Zhdanov, A. V. Kiselev, and L. F. Pavlova, *Kinetics and Catalysis (U.S.S.R.)*, 1962, **3**, 391.
21. J. H. Hildebrand and R. L. Scott, "The Solubility of Non-Electrolytes", Reinhold, New York, 1950, 3rd edition.
22. J. F. Goodman and S. J. Gregg, *J. Chem. Soc.*, 1959, 694.
23. R. S. Hansen, R. D. Hansen, and R. P. Craig, *J. Phys. Chem.*, 1953, **57**, 215.
24. J. L. Morrison and D. M. Miller, *Canad. J. Chem.*, 1955, **33**, 330.
25. E. R. Linner and A. P. Williams, *J. Phys. Chem.*, 1950, **54**, 605.
26. A. V. Kiselev and K. Shcherbakova, *Acta Physicochim. U.R.S.S.*, 1946, **21**, 539.
27. A. V. Kiselev, T. A. Worms, V. V. Kiseleva, and N. A. Shtovish, *Zhur. fiz. Khim.*, 1945, **19**, 83.
28. D. H. Everett, "The Structure and Properties of Porous Materials" (Everett and Stone, ed.), Butterworths, London, 1958, p. 95.
29. W. F. Wolff, *J. Phys. Chem.*, 1958, **62**, 829.
30. L. G. Ganichenko, V. F. Kiselev, and K. G. Krasil'nikov, *Doklady Akad. Nauk S.S.S.R.*, 1959, **125**, 1277.
31. J. J. Kipling and R. B. Wilson, *J. Appl. Chem.*, 1960, **10**, 109.
32. A. L. Elder and R. A. Springer, *J. Phys. Chem.*, 1940, **44**, 943.
33. J. J. Kipling and E. H. M. Wright, *J. Chem. Soc.*, 1964, 3535.

34. W. R. Smith, "Encyclopedia of Chemical Technology" (Kirk and Othmer, ed.), Interscience, New York, 1949, Vol. III, p. 24.
35. M. L. Studebaker, *Rubber Chem. Technol.*, 1957, **30**, 1400.
36. W. F. K. Wynne-Jones, "The Structure and Properties of Porous Materials" (Everett and Stone, ed.), Butterworths, London, 1958, p. 35.
37. C. G. Gasser and J. J. Kipling, *J. Phys. Chem.*, 1960, **64**, 710.
38. M. Hille and C. Jentsch, *Z. Krist.*, 1963, **118**, 283.
39. A. Blackburn and J. J. Kipling, *J. Chem. Soc.*, 1954, 3819.
40. A. Blackburn, J. J. Kipling, and D. A. Tester, *J. Chem. Soc.*, 1957, 2373.
41. D. C. Jones and G. S. Mill, *J. Chem. Soc.*, 1957, 213.
42. J. J. Kipling and D. B. Peakall, *J. Chem. Soc.*, 1957, 834.
43. V. S. Komarov and N. F. Ermolenko, *Russ. J. Phys. Chem.*, 1961, **35**, 5.
44. G. Delmas and D. Patterson, *Off. Dig. Fed. Paint Varn. Prod. Cl.*, 1959, **31**, 1129.
45. G. A. H. Elton, *J. Chem. Soc.*, 1952, 1955.
46. E. Heymann and E. Boye, *Z. phys. Chem.*, 1930, A **150**, 219.
47. E. R. Linner and R. A. Gortner, *J. Phys. Chem.*, 1935, **39**, 35.
48. I. M. Kolthoff and E. van der Groot, *Rec. Trav. chim.*, 1929, **48**, 265.
49. P. Gupta and P. De, *J. Indian Chem. Soc.*, 1946, **23**, 353.
50. W. A. Patrick and N. F. Eberman, *J. Phys. Chem.*, 1925, **29**, 220.
51. A. C. Chatterji and R. D. Srivastava, *J. Indian Chem. Soc.*, 1952, **29**, 325.
52. J. Traube, *Annalen*, 1891, **265**, 27.
53. H. Freundlich, "Colloid and Capillary Chemistry", Methuen, London, 1926.
54. L. Lepin, *Z. phys. Chem.*, 1931, A **155**, 109.
55. E. Landt and W. Knop, *Z. phys. Chem.*, 1932, A **162**, 331.
56. S. Claesson, *Arkiv. Kemi. Min. Geol.*, 1946, **23**, No. 1.
57. A. Baum and E. Broda, *Trans. Faraday Soc.*, 1938, **34**, 797.
58. E. Miculicich, *Arch. expt. Path. Pharmak.*, 1933, **172**, 373.
59. B. Tamamushi, *Bull. Chem. Soc. Japan*, 1932, **7**, 168.
60. A. Blackburn and J. J. Kipling, *J. Chem. Soc.*, 1955, 1493.
61. F. H. M. Nestler and H. G. Cassidy, *J. Amer. Chem. Soc.*, 1930, **72**, 680.
62. J. J. Kipling, *J. Chem. Soc.*, 1952, 2858.
63. I. Langmuir, *J. Amer. Chem. Soc.*, 1917, **39**, 1848.
64. B. Nekrassow, *Z. phys. Chem.*, 1928, **136**, 379.
65. H. N. Holmes and J. B. McKelvey, *J. Phys. Chem.*, 1928, **32**, 1522.
66. F. S. Bartell and Y. Fu, *J. Phys. Chem.*, 1928, **32**, 676.
67. B. Iliin, *Z. phys. Chem.*, 1931, A **155**, 403.
68. B. Bruns, *Kolloid Z.*, 1931, **54**, 33.
69. E. Heymann and E. Boye, *Kolloid Z.*, 1932, **59**, 153.
70. D. J. Crisp, *J. Colloid Sci.*, 1956, **11**, 356.
71. R. S. Hansen and R. P. Craig, *J. Phys. Chem.*, 1954, **59**, 211.
72. R. S. Hansen and R. D. Hansen, *J. Phys. Chem.*, 1955, **59**, 496.
73. R. S. Hansen, Y. Fu, and F. E. Bartell, *J. Phys. Chem.*, 1949, **53**, 769.
74. B. Nekrassow, *Z. phys. Chem.*, 1928, **136**, 18.
75. R. U. Lemieux and J. L. Morrison, *Canad. J. Res.*, 1947, B **25**, 440.
76. L. H. Klemm, D. Reed, L. A. Miller, and B. T. Ho, *J. Org. Chem.*, 1959, **24**, 1468.
77. L. Zechmeister, *Ann. N.Y. Acad. Sci.*, 1948, **49**, 220.
78. A. M. Rosen, *Compt. rend. Acad. Sci. U.R.S.S.*, 1943, **41**, 296.

CHAPTER 11

Adsorption at the Liquid–Vapour Interface

INTRODUCTION

The nature of the liquid–vapour interface can be regarded as proper to an extension of the theory of liquids. In this respect it has been considered in terms of experimental data on surface tension.[1] It is proposed, in this chapter, to deal only with sufficient aspects of the interface to show the extent to which it resembles the liquid–solid interface. Insoluble films (including protein films formed by adsorption from solution and subsequent denaturation) are not considered.

Direct Experimental Evidence

Addition of a second component to a pure solvent results in a change of surface tension. For almost all mixtures it can be shown that the surface layer has a different composition from that of the bulk of the mixture. One component is then said to be "adsorbed" at the surface. Adsorption (or surface excess) in this sense has, however, been defined in several ways, as is discussed below.

Direct experimental evidence of such adsorption is not easily obtained, because it is difficult to separate the thin surface layer from the bulk liquid sufficiently well to measure differences in composition. Two striking experiments were carried out by McBain which are impressive, not for their absolute accuracy, but in the ingenuity used in separating the surface from all but a small quantity of the bulk liquid. These are described in Chapter 2.

For dilute aqueous solutions of hydrocinnamic acid (1·5 g./l.), the microtome and barrier methods both gave surface excesses of about $5·2 \times 10^{-8}$ g./sq. cm. (3·5 micromoles/sq. m.), compared with $5·9–6·3 \times 10^{-8}$ g./sq. cm. calculated from surface tension data by means of the Gibbs equation. (The value obtained by the barrier method depended markedly on the age of the system, and the maximum value is given.) Slightly less close agreement was obtained for solutions of lauryl sulphonic acid. The "foam" method (Chapter 2) also gave values for the surface excess of pelargonic acid ($C_9H_{18}O_2$) in aqueous solutions which were in reasonable agreement with those calculated from the Gibbs equation.

191

DEFINITIONS OF ADSORPTION OR SURFACE EXCESS

As direct measurement of "adsorption" at the liquid–vapour interface is difficult, it has been usual to calculate the required value from an accurate measurement of surface tension. This is done by means of the Gibbs "adsorption" equation, which can be quoted in several forms. To understand the use of this equation, it is necessary to distinguish between the surface concentrations of the individual components and their surface excess, the latter being the extent by which the surface concentration of a given component exceeds its concentration in the bulk liquid. The surface concentration corresponds to the terms n_1^s and n_2^s of equation (3.3), and the surface excess to the term $n_0 \Delta x/m$.

Surface excesses can be defined in several ways (cf. $n_0 \Delta x/m$ and $w_0 \Delta c/m$) and have correspondingly different values. For comparison of different interfaces it is therefore particularly important to ensure that the adsorption terms have been defined and the results calculated in the same way.

The Gibbs equation is often quoted in an approximate form which is only appropriate for dilute solutions (see p. 195). A general and precise form is:

$$-d\gamma = \sum_i \Gamma_i d\mu_i = RT \sum_i \Gamma_i d\ln a_i, \qquad (11.1)$$

where γ is the surface tension of a solution in which the chemical potential and activity of component i are μ_i and a_i, respectively, and Γ_i is the surface *concentration* of that component. Thus for a two-component mixture:

$$-d\gamma = \Gamma_1 d\mu_1 + \Gamma_2 d\mu_2, \qquad (11.2)$$

whence, from the Gibbs–Duhem equation $(x_1 d\mu_1 + x_2 d\mu_2 = 0)$,

$$d\gamma = \Gamma_1 \frac{x_2 d\mu_2}{x_1} - \Gamma_2 d\mu_2,$$

or

$$\frac{x_1 d\gamma}{d\mu_2} = \Gamma_1 x_2 - \Gamma_2 x_1. \qquad (11.3)$$

Correspondingly,

$$\frac{x_2 d\gamma}{d\mu_1} = \Gamma_2 x_1 - \Gamma_1 x_2. \qquad (11.4)$$

Equation (11.3) is similar in form to equation (3.3), which suggests that the term $\frac{x_2 d\gamma}{d\mu_1}$ has the significance of surface excess; Γ_1 and Γ_2 correspond to n_1^s and n_2^s, respectively.

The different ways in which surface excess has been defined or used have been made very clear by Guggenheim and Adam,[2] whose nomenclature is followed here. The results of calculating surface excess according to the various definitions are illustrated very clearly by Adamson.[3] In each case the surface excess of component i is in moles per unit area.

$\Gamma_i^{(N)}$. For comparison of different interfaces, the $\Gamma_i^{(N)}$ convention is probably the most useful. $\Gamma_i^{(N)}$ is the excess of component i in the surface layer of unit area compared with the amount present in the quantity of bulk liquid which contains the same total number of moles of all species. For a two-component liquid, this means that:

$$\Gamma_2^{(N)} = \Gamma_2 - x_2(\Gamma_1 + \Gamma_2),$$

or

$$\Gamma_2^{(N)} = \Gamma_2 x_1 - \Gamma_1 x_2. \tag{11.5}$$

Hence from equation (11.4),

$$\Gamma_2^{(N)} = \frac{x_2 \, d\gamma}{d\mu_1},$$

or

$$\Gamma_2^{(N)} = -\frac{x_1 \, d\gamma}{d\mu_2}. \tag{11.6}$$

Equation (11.6) can also be written as

$$\Gamma_2^{(N)} = \frac{-x_1 \, d\gamma}{RT \, d\ln a_2}, \tag{11.7}$$

where a_2 is the activity of component 2 in the liquid phase. If the vapours of the two components obey the perfect gas laws, it is convenient, for experimental purposes, to write

$$\Gamma_2^{(N} = \frac{-x_1 \, d\gamma}{RT \, d\ln p_2}. \tag{11.8}$$

Alternative Definitions of Surface Excess

$\Gamma_i^{(1)}$. The surface layer can be defined so that $\Gamma_i^{(1)}$ (the surface excess of the "solvent") is zero. This definition is of most value for dilute solutions. Then $\Gamma_i^{(1)}$ is the excess of component i in surface layer of unit area compared with the amount present in that quantity of the bulk liquid which contains the same number of moles of component 1. For a two-component mixture this means that:

$$\Gamma_2^{(1)} = \Gamma_2 - \frac{x_2}{x_1} \Gamma_1,$$

or

$$\Gamma_2^{(1)} = \frac{1}{x_1} (\Gamma_2 x_1 - \Gamma_1 x_2). \tag{11.9}$$

Thus

$$\Gamma_2^{(1)} = -\frac{d\gamma}{RT \, d\ln a_2}. \tag{11.10}$$

$\Gamma_i^{(i)}$. The same convention can be applied with the surface excess of any other component defined as zero, e.g. $\Gamma_2^{(2)} = 0$.

$\Gamma_i^{(M)}$. In this definition, the portions of the surface layer and of the bulk liquid which are compared, have the same mass. Thus, for a two-component mixture:

$$\Gamma_2^{(M)} = \Gamma_2 - \frac{\Gamma_1 M_1 + \Gamma_2 M_2}{x_1 M_1 + x_2 M_2} x_2,$$

or

$$\Gamma_2^{(M)} = \frac{M_1}{\overline{M}} (\Gamma_2 x_1 - \Gamma_1 x_2), \tag{11.11}$$

where M_1 and M_2 are the molecular weights of components 1 and 2, respectively, and $\overline{M}$ is the mean molecular weight of the solution. Then

$$\Gamma_2^{(M)} = -\frac{M_1}{\overline{M}} \frac{x_1 \, d\gamma}{RT \, d\ln a_2}. \tag{11.12}$$

$\Gamma_i^{(v)}$. In this definition, the portions of surface layer and of the bulk liquid which are compared have the same volume. Thus, for a two-component liquid:

$$\Gamma_2^{(v)} = -\frac{V_1}{\overline{V}} \frac{x_1 \, d\gamma}{RT \, d\ln a_2}, \tag{11.13}$$

where V_1 and $\overline{V}$ are the partial molar volume of component 1 and the mean partial molar volume of the solution, respectively.

The relationship between the surface excesses calculated according to the different conventions is:

$$\Gamma_2^{(N)} = x_1 \Gamma_2^{(1)} = \frac{\overline{M}}{M_1} \Gamma_2^{(M)} = \frac{\overline{V}}{V_1} \Gamma_2^{(v)}. \tag{11.14}$$

For comparison of the liquid–vapour and liquid–solid interfaces, we can write:

$$-\Gamma_2^{(N)} = \Gamma_1^{(N)} = \frac{n_0 \Delta x}{Sm} \tag{11.15}$$

where S is the specific surface area of the solid.

(Some confusion may be avoided if two different past practices are noted. In adsorption at the liquid–solid interface, it has been usual to plot preferential adsorption of component 1. Many authors in dealing with the liquid phase now regard the "solvent" as component 1, and plot surface excess of component 2. No difficulty should arise if the axes of diagrams are labelled with the name of the relevant component.)

Usage for Dilute Solutions

From equations (11.5, 11.9, and 11.11) it can be seen that for dilute solutions (x_2 small), the surface excess approximates to the surface concentration.

This is analogous to the situation for the liquid–solid interface described in equation (3.9). In such cases, the activity coefficient of component 2 tends to unity, and the Gibbs equation is then quoted in the approximate form:

$$\Gamma = -\frac{c \, d\gamma}{RT \, dc},$$ (11.16)

in which Γ tends to be used indiscriminately for surface excess or surface concentration. It is doubtless in this way that the two terms come to be confused when more concentrated solutions are being considered. Sufficient confusion has existed in the past for the nature of surface excess (as given by the Gibbs equation) to be specially emphasized.[4]

COMPARISON OF SURFACE EXCESS AT DIFFERENT INTERFACES

The above treatment emphasizes that adsorption from a two-component solution at the liquid–solid interface is analogous to adsorption at the corresponding liquid–vapour interface. Adsorption at these two interfaces can be compared by means of equation (11.15). As yet, it has been little used for completely miscible liquids, partly because the similarity between the interfaces has not been sufficiently recognized, but also because results for adsorption on solids have been quoted without reference to the specific surface area; in much of the earlier work this was probably not known. A number of well-characterized adsorbents have recently become available, and it should now be possible to make considerable progress in comparing the two interfaces. For this purpose it is very convenient to express Γ, for both types of interface, in μmoles/sq. m. The equivalent unit of 10^{-10} moles/sq. cm. has sometimes been used, and it has been suggested[5] that it be called "the Gibbs", with the symbol, G.

Some preliminary results have yielded interesting qualitative conclusions.[6] For liquid systems showing complete miscibility, the component of lower surface tension is preferentially adsorbed at the liquid–vapour interface. Adsorption by a solid from the same mixture may be of the same (Fig. 11.1) or of opposite preference (Fig. 11.2). The magnitude of the surface excess is comparable for the two interfaces (Fig. 11.2). It is generally greater at the liquid–solid interface, but a ten-fold difference is extreme. Thus the high adsorptive capacity of solid adsorbents results primarily from their high specific surface area rather than from a very high degree of preferential adsorption of one component.

Adsorption at the liquid–vapour interface, governed mainly by the relative surface tensions of the two components, is normally U-shaped. For the liquid–solid interface, however, it may be U-shaped or S-shaped, according to factors which are discussed in Chapter 10. Some of these may also apply, to a minor extent, to adsorption at the liquid–vapour interface. Thus for aqueous solutions, King and Wampler found that dicarboxylic

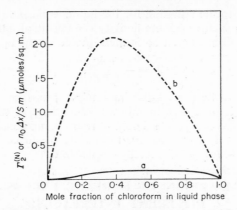

FIG. 11.1. Adsorption from mixtures of benzene and chloroform[6]: (a) at the liquid–vapour interface (25°C), (b) on boehmite (20°C). (Reproduced from the "Proceedings of the Third International Congress of Surface Activity", Verlag der Universitätsdruckerei, Mainz.)

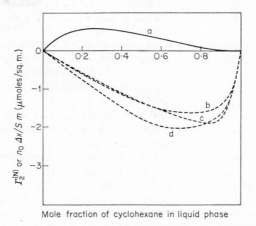

FIG. 11.2. Adsorption from mixtures of benzene and cyclohexane at 20°C: (a) at the liquid–vapour interface, (b) on Spheron 6, (c) on charcoal, (d) on boehmite.[6] (Reproduced from the "Proceedings of the Third International Congress of Surface Activity", Verlag der Universitätsdruckerei, Mainz.)

acids were less strongly adsorbed than the corresponding monocarboxylic acids, and attributed this to the greater attraction of the water for two carboxylic groups than one.[7] Correspondingly, the esters are more strongly adsorbed than the acids, because the ester group is less polar than the carboxyl group. Increasing length of aliphatic chain normally causes increasing adsorption (cf. Traube's Rule, below), though in an irregular way for the dicarboxylic acids. Polar substituents also affect the strength of adsorption, as measured by the lowering of surface tension at a given molar concentration.

By using values of $\Gamma^{(v)}$, Bartell and Benner have also compared surface excesses for different interfaces in the same units.[8] They showed that adsorption from the system iso-amyl alcohol–iso-octane gave very similar isotherms for the liquid–solid and the liquid–water interfaces, as was also the case for the system cyclohexanol–iso-octane. They also compared adsorption at the two interfaces by using the "mixture law". This is based on the assumption that, in the absence of adsorption, the surface tension (or interfacial tension) of a liquid mixture would vary linearly with the volume composition.[9] (The assumption would not apply to liquids showing association or dissociation.) The difference between the observed surface tension and that calculated according to the linear relationship can then be used to calculate the surface excess. In this method it is given in moles/c.c. It has therefore been used in conjunction with Γ (in moles/sq. cm.) to calculate the thickness of the adsorbed layer. The validity of this last calculation, or the further assumptions underlying it, have not yet been carefully scrutinized.

A theoretical basis for comparing adsorption at different interfaces for ideal or regular solutions is outlined in Chapter 14. Other methods for partially miscible liquids are considered later in this chapter.

NATURE OF THE ADSORBED LAYER

Individual Isotherms

Once the surface excess has been calculated, it is possible in principle to obtain the actual surface concentration for each component by using equation (11.5) or an analogous equation. This process corresponds to resolving the composite isotherm for adsorption at the liquid–solid interface into the individual isotherms. The same kind of assumptions have to be made: as to the orientation of the adsorbed molecules at the interface, the area which they occupy, and whether the adsorbed layer is monomolecular or multimolecular. The result of the analysis is therefore subject to the limitations discussed in Chapter 4. Moreover, for this interface, there is no subsidiary evidence as to the likely orientation of the molecules such as is given for the liquid–solid interface by considering adsorption of the individual vapours by the solids. The value of the analysis therefore depends on the soundness with which orientation and molecular areas are judged.

Thickness of the Adsorbed Layer

Although there has been some controversy concerning the thickness of the adsorbed layer at the liquid–solid interface, it seems generally to have been assumed that the adsorbate at the liquid–vapour interface is confined to the thickness of one molecule. This, as Guggenheim and Adam[2] pointed out, is the simplest assumption to make. Direct evidence is difficult to obtain, and the general assumption appears to be accepted mainly because it gives results compatible with values of Γ derived from the Gibbs equation. Moreover, the values of Γ are, for the majority of systems, very much smaller than would require multilayer adsorption to be postulated.

The test proposed by Schay and Nagy was originally applied to the liquid–solid interface, but can in principle be applied equally well to the liquid–vapour (or liquid–liquid) interface. In practice, however, few isotherms are available which have a substantial linear section. One such isotherm, for the system water–ethyl alcohol,[2] has been treated in this way,[10] and the adsorbate has been shown to be confined to a monolayer; the same conclusion holds for methyl alcohol–water. The analysis depends on the choice of molecular areas for the two components (10 sq. Å for water, which is a generally accepted value, and 18·4 sq. Å for ethyl alcohol, which corresponds to the orientation with the major axis of the molecule perpendicular to the surface and the —OH groups towards the solution). This result appears to dispose of earlier suggestions that the surface layer is more complex than this,[11,12] but it is remarkable that another monolayer model[13] has given satisfactory results with quite different molecular areas: 17 sq. Å/molecule for water, and 43 and 75 sq. Å/molecule for ethyl alcohol at low and high concentrations respectively.

A treatment put forward by Schofield and Rideal for dilute solutions[12] adds further support to the monolayer hypothesis. Following Traube, they defined a quantity F:

$$F = \gamma_0 - \gamma, \qquad (11.17)$$

where γ_0 is the surface tension of the pure solvent and γ that of the solution, and found that a number of data fitted the equation

$$F(A - B) = kTx, \qquad (11.18)$$

where A is the area available to each solute molecule in the surface layer, B the minimum area which each solute molecule can occupy, k is the Boltzmann constant and $1/x$ is a measure of the molecular cohesion. If F is regarded as a measure of surface force, this equation is an analogue, for an imperfect two-dimensional film, of Amagat's equation for an imperfect three-dimensional gas.

Application of this equation to data obtained by Szyszkowski gave the following results for dilute aqueous solutions:

	B (sq. Å)	x	$\dfrac{1}{x}$
n-butyric acid	24·3	0·73	1·37
n-valeric acid	24·3	0·63	1·59
n-caproic acid	24·3	0·43	2·32
iso-butyric acid	25·1	0·78	1·28
iso-valeric acid	25·1	0·68	1·47
iso-caproic acid	25·1	0·48	2·08

The significant features are: (i) that each set of acids gives a constant surface area per molecule (24·3 compared with 20·5 sq. Å per molecule for closest packing for the n-acids), which implies that all the acids have the same molecular orientation, with an area close to that expected for formation of a complete monolayer; (ii) that the intermolecular forces are less between the iso- than between the n-acids, a necessary effect of the methyl substituent with the orientation postulated; (iii) that the intermolecular forces increase with chain-length for both series, which is also required with this orientation.

A limiting case might be expected at concentrations near solubility limits. Ellison and Zisman found that adsorption of fluorinated esters at the hexadecane–air interface could be interpreted in terms of the formation of an almost complete monolayer at concentrations close to the solubility limit.[14] In general, the effect of raising the concentration of solute in the liquid phase is to displace solvent molecules from the adsorbed phase. This can be expressed as increasing the probability of collision between molecules in the adsorbed phase. If this becomes high and if the cohesional energy between the solute molecules is also high, nucleation and subsequent crystallization result. If the cohesional energy is small (as is notably the case with fluorinated compounds) a stable monolayer is more likely to be formed.

For a few systems, multilayer adsorption has to be postulated. The first is cyclohexane–aniline at 40°C, for which only approximate values of Γ, obtained by extrapolation, are available. If, as seems likely, the adsorbed layer exceeds a monolayer,[6] this system would be analogous to those giving multilayer adsorption at the liquid–solid interface in being close to phase separation. The critical solution temperature for the system is 31°C.

Three further systems in which a monolayer is exceeded are water–acetone, water–dioxan, and water–triethylamine.[15] Breitenbach and Edelhauser point out that the second components are all weakly surface-active. The high concentration at the surface appears to be associated with a tendency towards phase separation. This, however, is not the only factor, for the adsorbed layer appears to be confined to a monolayer for the system water–n-butyl alcohol at 30°C up to the point at which phase separation occurs.[15]

The distinction between monolayer and multilayer adsorption is one which cannot be drawn rigidly in most cases. It can only be applied precisely to systems in which the molecules of the two components have the same thickness perpendicular to the surface. Few systems fulfil this criterion exactly, though many which have been investigated do so to a close degree of approximation. Disparities in molecular size can, however, be considerable. Thus Fig. 11.3 shows the relative sizes of molecules of water and n-octyl alcohol. If the latter is oriented with the major axis perpendicular to the surface, its thickness is about that of four water molecules. The effective thickness of the surface layer (in terms of competition between the two components for positions at the surface) then depends on whether the hydrocarbon chain is immersed in the bulk liquid (Fig. 11.3a) or not (Fig. 11.3b). In either case thermal motion of the molecules makes it impossible to define

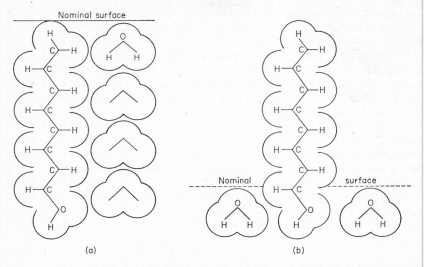

FIG. 11.3. Relative sizes of molecules of water and n-octyl alcohol.[20] (Reproduced from the *Journal of Colloid Science*.)

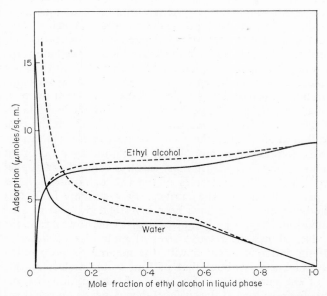

FIG. 11.4. Individual isotherms for adsorption of ethyl alcohol and water at the liquid–vapour interface (25°c); broken curves calculated for a double layer of water molecules.[20] (Reproduced from the *Journal of Colloid Science*.)

the position of the "nominal" surface exactly. The distinction between a monolayer and a multilayer of adsorbate can, therefore, not be made rigidly in this case or, indeed, in most cases.

Some Individual Isotherms

By using equations analogous to (3.2) and (4.19) Guggenheim and Adam calculated the individual isotherms for the system ethyl alcohol–water.[2] The equations were (11.5) and (11.19):

$$\Gamma_1 A_1 + \Gamma_2 A_2 = 1, \tag{11.19}$$

where A_1 and A_2 are the partial molar areas of components 1 and 2, respectively. The results have recently been re-calculated,[10, 20] and are given in Fig. 11.4 for comparison with those for a typical non-polar system, chloro-

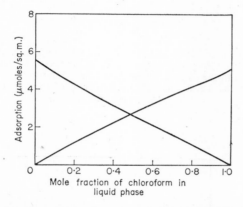

Fig. 11.5. Individual isotherms for adsorption of chloroform and carbon tetrachloride at the liquid–vapour interface[6] (18°c). (Reproduced from the "Proceedings of the Third International Congress of Surface Activity", Verlag der Universitätsdruckerei, Mainz.)

form–carbon tetrachloride (Fig. 11.5). As the latter two molecules are almost spherical, there is little doubt as to the area which they occupy in the surface, and consequently as to the shapes of the individual isotherms. It is clear that the surface layer differs very little in composition from that of the bulk liquid. This "composite" isotherm, which shows the quite definite preferential adsorption of chloroform, is thus a very sensitive measure of adsorption at this interface. The broken curves in Fig. 11.4 show the effect of assuming that the oriented ethyl alcohol molecule is as thick as two water molecules. The isotherms are not greatly affected by this assumption for the system in question because the preferential adsorption of the ethyl alcohol is so strong.

The effect of alternative molecular orientations for one component is shown in Fig. 11.6 for the system benzene–carbon tetrachloride and for both

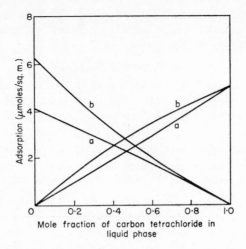

FIG. 11.6. Individual isotherms for adsorption of benzene and carbon tetrachloride at the liquid–vapour interface (50°C), calculated for adsorption with the plane of the benzene molecule: (a) parallel to the surface, (b) perpendicular to the surface.[6] (Reproduced from the "Proceedings of the Third International Congress of Surface Activity", Verlag der Universitätsdruckerei, Mainz.)

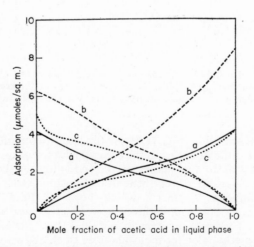

FIG. 11.7. Individual isotherms for adsorption of benzene and acetic acid at the liquid–vapour interface (35°C) with the major axes of both molecules: (a) parallel to the surface, (b) perpendicular to the surface.[6] Curves marked (c) show the isotherms for adsorption on charcoal at 20°C. (Reproduced from the "Proceedings of the Third International Congress of Surface Activity", Verlag der Universitätsdruckerei, Mainz.)

components in Fig. 11.7 for benzene–acetic acid. The benzene molecule has
extreme orientations with its plane parallel to and perpendicular to the
surface; the acetic acid molecule (taken as being the dimer in the mixtures
with benzene) may similarly have its major axis parallel to or perpendicular
to the surface in its extreme positions. It may be significant that the individual
isotherms for the benzene–acetic acid system closely resemble those for
adsorption on charcoal if the major axes are assumed to be parallel to the
surface in all four cases (Fig. 11.7c). Similarly it seems likely that acetic
acid and its homologues tend towards this orientation from a random
one at the nitromethane–air interface.[16] Data obtained for the liquid–solid
interface may therefore sometimes be useful as guidance in the interpretation
of data for the liquid–vapour interface. Caution is needed, however, as there
is no necessity for the orientations to be the same at the two interfaces;
indeed, a polar substance can be oriented in different ways at two solid
surfaces, particularly if one is highly polar and the other virtually non-polar.

Proposals have also been made to calculate the actual concentrations of
each component at the interface from $\Gamma^{(v)}$. Bartell and Benner[8] used the
equation:

$$Z_2 = \Gamma_2^{(v)} + \frac{V_2 N_2}{mV}, \qquad (11.20)$$

where Z_2 is the number of molecules of solute (component 2) present in
1 sq. cm. of interface, V_2 is the molar volume of the solute, m its molar area,
N_2 its mole fraction in the solution, and V the molar volume of the solution.
For this equation to be used, an estimate must be made of the molecular
area of the solute in the interface (or alternatively of the thickness of the
adsorbed layer). This procedure is mathematically not identical with that
described above, based on values of $\Gamma^{(N)}$, in which the molecular areas of
both components are used. In principle, when the values of $\Gamma^{(v)}$ are used
it would seem desirable to estimate Z_1 as well as Z_2 if the results are to be
as satisfactory as those based on values of $\Gamma^{(N)}$. In practice, however, the use
of $\Gamma^{(v)}$ seems to have been restricted to cases in which preferential adsorption
of one component is very strong. For such systems, the effectiveness of the
calculation probably then depends on how well the molecular area is estimated. The results of making such an estimate in three different ways are
illustrated by Bartell and Benner. For a strongly adsorbed solute, the
differences may be small at low mole fractions, but become important at
high mole fractions.

If the thickness of the adsorbed layer is used, the film is assumed to be
unimolecular with respect to the component having the longer molecule;
it may thus be more than one molecule thick with respect to the shorter
molecule. By contrast, in the use of $\Gamma^{(N)}$ values, the adsorbed layer is normally
assumed to be unimolecular with respect to both components; alternative
assumptions can, however, be specifically used in calculating the individual
isotherms.

One aspect of these calculations is brought out by Jones and Saunders, who point out that the value of the absolute adsorption from organic solvents is much more controlled by the bulk concentration (as opposed to the surface excess) than is adsorption from aqueous solutions.[16] This can alternatively be expressed by saying that surface-active solutes are generally more strongly adsorbed at this interface from water than from other solvents.

Partially Miscible Liquids

The usual treatment of partially miscible liquids (or of solutions of a solid) is to compare the adsorbed layer with that formed by an insoluble amphipathic substance on an aqueous substrate. Such a comparison has led first to a consideration of the orientation of the adsorbed solute. The surface excesses of aqueous solutions of hydroquinone, resorcinol, pyrocatechol, and pyrogallol are consistent with the formation of monolayers, with less than the closest packing, the orientation being determined by attraction of the hydroxyl groups towards water in the bulk phase and corresponding rejection of the benzene rings.[17]

Solutions of n-butyl alcohol in water have been considered in more detail. Harkins and Wampler[18] used the Gibbs equation to calculate the surface excess, expressed in terms of $\Gamma_2^{(1)}$. By adding the number of butyl alcohol molecules which would be present if there were no surface excess, they obtained the total number present per unit area of surface. From the number of molecules present in unit area could be calculated the area within which one molecule of butyl alcohol could be found. (This phrase is used rather than the more conventional phrase "area occupied per molecule", because part of this area is presumably occupied by water molecules.) The number of butyl alcohol molecules present in unit area of surface if there were no adsorption was assumed to be the two-thirds power of the number of alcohol molecules present in 1 c.c. of solution. The results are shown in Table 11.1.

TABLE 11.1

Molality	Surface excess $\Gamma_2^{(1)}$ (molecules × 10^{14}/sq. cm.)	Addition for BuOH present if $\Gamma = 0$ (Harkins) (molecules × 10^{14}/sq. cm.)	Total surface concentration of BuOH (molecules × 10^{14}/sq. cm.)	Area per molecule (sq. Å)	Correction according to Butler (molecules × 10^{14}/sq. cm.)	Corrected surface concentration of BuOH (molecules × 10^{14}/sq. cm.)	Corrected area per molecule (sq. Å)
0·01320	0·76	0·04	0·80	120	0·002	0·76	120
0·0264	1·31	0·06	1·37	73	0·004	1·31	73
0·0536	2·14	0·10	2·24	45	0·006	2·15	47
0·1050	2·86	0·15	3·01	33	0·009	2·87	35
0·2110	3·19	0·23	3·42	29·2	0·016	3·21	31·2
0·4330	3·45	0·37	3·82	26·2	0·028	3·48	29·8
0·8540	3·65	0·56	4·22	23·7	0·050	3·70	27·0

Butler claimed that this assumption is, in general, an overestimate.[19] From equation (11.9):

$$x_1 \Gamma_2^{(1)} = x_1 \Gamma_2 - (1-x_1)\Gamma_1,$$

or

$$\Gamma_2^{(1)} = \Gamma_2 - \frac{1-x_1}{x_1}\Gamma_1,$$

whence

$$\Gamma_2 = \Gamma_2^{(1)} + \frac{1-x_1}{x_1}\Gamma_1. \tag{11.21}$$

The correction term is thus $\frac{1-x_1}{x_1}\Gamma_1$, which Butler believed to be negligible for the concentrations concerned. An approximate estimate of these corrections is given in Table 11.1; they are based on the assumption that, at these low concentrations of butyl alcohol, Γ_2 is approximately equal to $\Gamma_2^{(1)}$, and can be used to calculate Γ_1 if it is assumed that adsorption is confined to a monomolecular layer and that suitable values (18·4 and 10·0 sq. Å) can be adopted for the areas occupied by the butyl alcohol and water molecules respectively in the surface film. The correction terms are seen to be quite small.

Whichever value is accepted for the area containing one butyl alcohol molecule, it decreases as the concentration of the solution rises. Data are available up to 0·8540 molal; the limit of solubility for the temperature concerned (25°C) is just over one molal, at which extrapolated values of about 23 or 27 sq. Å per molecule of butyl alcohol are obtained. These are compatible with a monomolecular adsorbed film in which the alcohol molecules are oriented with the major axes perpendicular to the surface; each molecule would be expected to occupy 18·4 sq. Å in this configuration.

(If the butyl alcohol molecules are oriented parallel to the surface they would occupy either 36·6 or 40·0 sq. Å per molecule, depending on which side of the hydrocarbon chain were presented to the surface. On this basis, the experimental result would correspond to multimolecular adsorption. Such an orientation, however, seems very unlikely to be adopted in aqueous solutions.)

An evaluation of Γ_1 and Γ_2 can be obtained much more simply and directly by using equations (11.9) and (11.19). Direct use of equation (11.9) avoids the need for corrections and is valid for all concentrations. Values of Γ_1 and Γ_2 for the n-butyl alcohol–water system obtained in this way are given in Table 11.2, and are plotted, in units of μmoles/sq. m., in Fig. 11.8. They show that the surface layer contains a high proportion of water. Even at the highest available concentration, the two components are present in approximately equimolecular proportions. Water molecules then occupy about 30% of the surface area, if the assumed molecular areas are correct.[20] This implies the formation of a surface film in which hydrogen-bonding through water molecules increases the stability of the molecular network.

TABLE 11.2

Molality	Mole fraction of butyl alcohol	Γ_2 (Molecules $\times 10^{14}$ of butyl alcohol per sq. cm.)	Γ_2 Butyl alcohol in surface layer	Γ_1 Water in surface layer
			(μmoles/sq.m.)	
0·0132	0·000238	0·76	1·26	14·3
0·0264	0·000475	1·32	2·19	12·6
0·0536	0·000967	2·15	3·57	10·0
0·1050	0·00189	2·85	4·73	7·88
0·2110	0·00379	3·21	5·33	6·78
0·4330	0·0076	3·52	5·85	5·85
0·8540	0·01515	3·70	6·15	5·30

It has been shown that the minimum areas occupied by many compounds at this interface are significantly greater than those of the corresponding insoluble compounds in close-packed films on aqueous substrates (aliphatic acids 30·5, aromatic acids 30·0, aliphatic and aromatic alcohols 28·5, aliphatic amines 30·0, aromatic amines, 28·0 sq. Å). As these higher values are not observed at the solution–mercury interface, the effect at the air

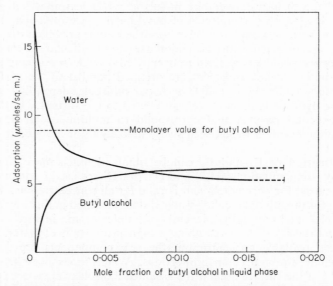

FIG. 11.8. Individual isotherms for adsorption of n-butyl alcohol and water at the liquid–vapour interface[20] (25°c). (Reproduced from the *Journal of Colloid Science*.)

interface is attributed to hydration of the solute.[21] For molecules like
p-methyl cyclohexanol, which are large enough for hydration not to affect
the packing in a surface layer, the same areas are observed at the two interfaces. Similar solvation of the aliphatic compounds occurs at the surfaces
of their solutions in formamide.[22]

It is interesting to note that butyl alcohol is more strongly adsorbed
from aqueous solution at the liquid–vapour or liquid–paraffin wax interface
than at interfaces with talc, stibnite, or graphite[23] (Fig. 11.9).

Fu and Bartell used existing surface tension data to calculate surface
excesses from the Gibbs equation for several aqueous solutions[24] (Fig. 11.10),
on the $\Gamma^{(V)}$ convention. Because dilute solutions are involved, these can
be considered as approximating to the individual isotherms for adsorption
of the solutes. Three of these isotherms (for solutions of butyl alcohol,

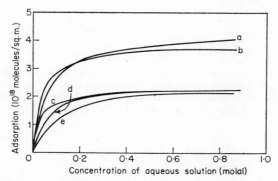

FIG. 11.9. Adsorption of n-butyl alcohol from aqueous solutions at interfaces with:
(a) paraffin, (b) air, (c) talc, (d) graphite, (e) stibnite.[23] (Reproduced with permission from
the *Journal of the American Chemical Society*.)

butyric acid, and resorcinol) reach a limiting value; this can be taken to
imply, though it does not prove, the existence of monolayer adsorption.
The remaining three do not reach a limiting value at the concentrations
studied, and as they resemble in shape Type II in Brunauer's classification
of isotherms for adsorption of single vapours on solids,[25] it was concluded
that multilayer adsorption occurred in this case. The general validity of such
assumptions is discussed in Chapter 5. In this case, the conclusion conflicts
with that obtained for valeric and caproic acids by Rideal and Schofield,
though it is supported by thermodynamic arguments which Bartell has put
forward (see below). Molecular areas for the adsorbate, derived by Bartell
and Benner's method,[8] do not seem too small, even at the highest concentrations quoted, to be compatible with monolayer adsorption.

There is nothing in these data to suggest that there is a significant difference
between adsorption of solid and liquid solutes at this interface, comparable
with that noted in Chapter 4 for the liquid–solid interface.

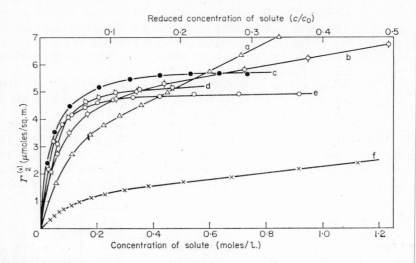

FIG. 11.10. Isotherms (various temperatures) showing surface excess of various solutes at the aqueous solution–vapour interface: (a) caproic acid, (b) valeric acid, (c) butyl alcohol, (d) phenol, (e) butyric acid, (f) resorcinol.[24] (Reproduced with the permission of the American Chemical Society from the *Journal of Physical Chemistry*.)

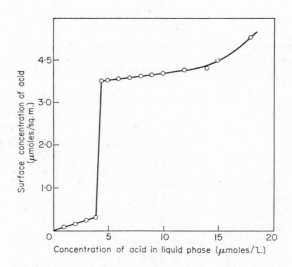

FIG. 11.11. Adsorption of n-undecanoic acid ($C_{11}H_{22}O_2$) at the air–water interface as a function of concentration.[26] (Reproduced from *Kolloid Zeitschrift* with the permission of Dr. Dietrich Steinkopff Verlag, Darmstadt.)

Very Dilute Solutions

Some remarkable results have been obtained by Dervichian[26] for the adsorption of the normal fatty acids C_6–C_{11} (caproic to undecanoic) at the air–water interface. The experiments have necessarily been carried out with very dilute solutions (of the order of 10^{-5} moles per litre). The Gibbs equation was used to calculate amounts adsorbed from surface tension measurements. The result for undecanoic acid is shown in Fig. 11.11.

The surface concentration rises linearly with concentration for the most dilute solutions. There is then an abrupt break in the curve, the adsorption rising more than ten-fold. Little further increase with concentration occurs until the region of micelle formation is reached. After the sudden increase in adsorption, the molecular area occupied in each case is in the region of 40–50 sq. Å.

Adsorption of Polymers at the Liquid–Vapour Interface

An interesting study of the adsorption of polyethylene glycols at the surface of aqueous solutions[27] shows the surface tension to vary linearly with the logarithm of concentration over the range of 10^{-3} to 10^2 g./l. From the Gibbs equation, it is therefore deduced that the surface excess of the polymer is constant over this wide range of concentration. Its magnitude, for four fractions of different molecular weight, corresponds to the formation of a complete monolayer with the major axes of the molecules parallel to the surface.

By contrast, polymethacrylic acid was found to form a surface layer on aqueous solutions with the thickness of a randomly coiled polymer molecule, which varied from 148 to 565 Å, corresponding to degrees of polymerization of 600 to 7800 (M.W. 51,600–670,800). This is the estimated mean diameter of randomly coiled polymeric molecules, the diameter varying according to the square root of the degree of polymerization. The surface activity was reduced to a very small value by as little as 20% ionization of the acid.[28]

Polymers at the water–air interface vary from being almost completely rigid (globular protein) through various degrees of flexibility (cellulose triacetate, polyvinyl alcohol) to having a random chain configuration (poly-n-butyl acrylate).[29] The detailed interpretation is a matter of controversy.[30]

Polymers are unusual in that they can, in some circumstances, raise instead of lower the surface tension of the solvent. This occurs with high molecular-weight fractions of polystyrene in tetralin (which is a "good" solvent for polystyrene), but not with low molecular weight fractions.[31] Similar results are found for block co-polymers of ethylene oxide and propylene oxide in the same solvent.[32] The theoretical aspect of this phenomenon is considered by Frisch and Simha.[33]

TRAUBE'S RULE

Traube observed a regularity in the lowering of the surface tension of water by members of the three homologous series of fatty acids, alcohols, and esters at low concentrations.[34] The generalization can be stated as follows[35]: In a homologous series of surface-active substances there is a constant ratio between the molar concentrations of two successive homologues giving rise to the same very small surface tension lowering in aqueous solutions. This law is usually known as Traube's Rule, although very similar observations had been put forward previously by Duclaux.[36] It is true only in the limiting case of extreme dilution, at which adsorbed films are considered to be in a state comparable with that of a two-dimensional ideal gas. The extension of the ideas inherent in Traube's Rule, especially to other interfaces, is considered in Chapter 10.

Langmuir[37] has given a physical significance to Traube's Rule. From the kinetic equilibrium between the surface layer and the interior of the solution it follows that

$$\frac{\Gamma}{c} = K\,e^{\lambda/RT}, \tag{11.22}$$

where λ is the work required to remove one mole of solute from the surface layer into the bulk of the solution, and K is a constant. The term $\frac{\Gamma}{c}$ can be replaced as follows. Szyszkowski found a relationship for the surface tensions of aqueous solutions of the lower fatty acids which can be expressed as[38]:

$$\frac{\gamma}{\gamma_0} = 1 - B\ln\left(\frac{c+A}{A}\right), \tag{11.23}$$

where A and B are constants.† Hence

$$d\gamma = -B\gamma_0\,\frac{dc}{c+A},$$

which, for dilute solutions, can be approximated to

$$d\gamma = B\gamma_0\,\frac{dc}{c}. \tag{11.24}$$

This is combined with the Gibbs equation to give:

$$\Gamma = \frac{B\gamma_0}{RT}\frac{c}{A}. \tag{11.25}$$

Hence, from equation (11.22)

$$\frac{B\gamma_0}{RT}\frac{1}{A} = K\,e^{\lambda/RT}. \tag{11.26}$$

† See Chapter 14 for a theoretical derivation of this equation.

Szyszkowski found B to be constant for all of the fatty acids which he studied. For any two members of a homologous series, therefore,

$$\frac{A_n}{A_{n+1}} = e^{(\lambda_{n+1} - \lambda_n)/RT}. \qquad (11.27)$$

Experimentally, it is found that A decreases in a constant ratio for each CH_2 group added to the hydrocarbon chain. This corresponds to a constant value for $e^{(\lambda_{n+1} - \lambda_n)}$, a term which is sometimes referred to as the Traube coefficient. For the fatty acid series, it corresponds to an increase in the work of desorption of 710 cal./mole for every CH_2 group.

Guastalla has used the Gibbs equation in the form[35]:

$$\left(\frac{\Gamma}{c}\right)_{limit} = \frac{1}{RT}\left(\frac{\pi}{c}\right)_0. \qquad (11.28)$$

to obtain the limiting ratio of $\dfrac{\Gamma}{c}$ at infinite dilution. π is the surface pressure

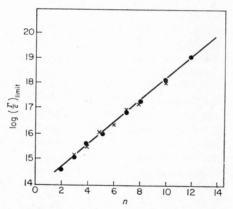

FIG. 11.12. Adsorption of fatty acids and alcohols at the water–air interface.[35] (Reproduced from the *Journal of Colloid Science*.)

(the difference between the surface tension of the solvent and that of the solution), and $\left(\dfrac{\pi}{c}\right)_0$ is the tangent of the curve of π against c at the origin.

Plots of $\log\left(\dfrac{\pi}{c}\right)_{limit}$ against n were obtained. These were linear for aqueous solutions of the fatty acids and alcohols from $n = 2$ to 12 (Fig. 11.12). The two curves give the same value for the Traube coefficient; hence for acids and alcohols having the same value of n, the average length of chain above the water surface is the same. These experiments give a lower value (2·65–2·70) for the Traube coefficient than had been obtained by Traube (whose value was 3) and in most subsequent investigations. It corresponds to a value of 390×10^{-16} ergs per CH_2 group (or 561 cal./mole) for the work of desorption.

From the constant work of desorption Langmuir concluded that for dilute solutions the hydrocarbon chain was adsorbed with its major axis parallel to the surface, but he added that the molecules would be oriented perpendicular to the surface when the solutions were concentrated. The parallel orientation is almost certain to be adopted at sufficiently low surface coverages (and thus at sufficiently low concentrations). Butler pointed out, however, that at the concentrations examined, the surface excess was too great to correspond to the parallel orientation.[20] The perpendicular orientation is thus the more likely one. Butler considered the major contribution to the free energy of adsorption to be the difference between the free energy of the pure solute in the liquid state and in dilute solution (in the standard state). This difference is related to the solubility (for dilute solutions) by the equation:

$$(\mu_2^0)_L - \mu_2^0 = RT \ln S, \qquad (11.29)$$

where S is the solubility. The limited data available support this view.

An alternative interpretation has been put forward by Aranow and Witten.[39] It is that the hydrocarbon chain projects into the vapour phase. For dilute solutions (and hence relatively dilute surface films), free rotation is possible at each C—C bond. The change in heat content on adsorption is assumed to depend chiefly on the polar group, and so to be constant for each member. Thus the change in free energy differs for each additional CH_2 group by an entropy term, $RT \ln 3$. On this basis, some deviation from the rule would be expected at the lower end of a homologous series. Deviations would also be expected in concentrated films as adjacent chains would have reduced freedom for free rotation. This view has been challenged, and the question as to whether the change in heat content is due mainly to an energy or to an entropy change is still under discussion.[40,41] Ward has shown that there are alternatives[42] to the cylindrical shape of the hydrocarbon chain, which Langmuir assumed to be adopted in the surface layer. A spheroidal shape appeared to be the most likely.

The differences between the above three views have not yet been resolved, but should be considered in relation to similar investigations of adsorption at the surface of paraffin wax. These give a higher value for the Traube coefficient, for the lower members, than is obtained at the liquid–vapour interface, but the same value for the higher members ($n > 7$).[43] Guastalla did not explain these results, which seem unlikely to be compatible with Aranow and Witten's views.

Traube's Rule has been considered for adsorption at other interfaces. The energy of adsorption per methylene group at the formamide–vapour interface[23] is 345 cal./mole. At this interface, the effect of the polar groups is not the same as at the water–vapour interface. Whereas acids, alcohols, and amines have approximately the same surface activity (at a given molar concentration) in aqueous solution, the activity decreases in the order alcohol > amine > acid in formamide solution, reflecting the differences in energy of interaction with the solvent.

Traube's Rule is usually considered in relation to homologous series of mono-functional compounds. Data have also been obtained for the first eight members of the di-carboxylic acid series.[44] In general, Traube's Rule is obeyed, and the calculated energy of desorption per CH_2 group is close to that found for the mono-carboxylic acids. This value was derived from acids with an even number of carbon atoms. Those with an odd number showed anomalously high surface activity. This was correlated with their high solubility in water, but was not further explained.

The quantitative treatment of Traube's Rule can be modified to apply to more concentrated solutions. This involves a recognition that the more concentrated surface films are no longer ideal, and allowance must be made for the interactions of adjacent molecules.[43]

ADSORPTION FROM TERNARY MIXTURES

Adsorption from solutions containing two solutes was investigated by Szyszkowski.[38] He postulated that two solutions, each containing one solute in a common solvent, and showing the same lowering of surface tension with respect to the pure solvent, should not alter in surface tension when mixed. He defined such solutions as *isotatic*, provided that they also obeyed the equation

$$\frac{\gamma}{\gamma_0} = 1 - B \ln\left(\frac{c+A}{A}\right), \tag{11.30}$$

where c is the concentration in moles per litre. It followed that:

$$\frac{c_1}{A_1} = \frac{c_2}{A_2} = \frac{c_{12}}{A_{12}}; \tag{11.31}$$

c_{12} is the concentration of the mixture (in moles per litre) and A_{12} is the capillary constant for the mixture required by equation (11.30). Hence, if a solution of concentration c_1 and a solution of concentration c_2 are mixed in the proportions $\alpha : (1-\alpha)$, then

$$A_{12} = \alpha A_1 + (1-\alpha)A_2. \tag{11.32}$$

Szyszkowski found this equation to apply over a considerable range of concentrations to mixtures of the lower fatty acids in aqueous solution.

A combination of equation (11.30) with the Gibbs equation gives:

$$\Gamma = \frac{\gamma_0 B}{RT} \cdot \frac{c}{c+A}. \tag{11.33}$$

This has the same form as can be obtained by applying Langmuir's equation to the adsorption of the solute.[45] The corresponding equation for two solutes can be used to show the degree of lowering of surface tension brought about by two solutes present in the same solution, e.g. for two alcohols in aqueous solution.[45]

Non-ideal behaviour has been found in adsorption from aqueous solutions containing both phenol and aniline, the lowering of surface tension being greater for the mixture than would be expected from the effects of the separate solutes. This can be attributed to increased adsorption arising from the attractive forces operating between the solutes which may act within a single layer;[46] for example, hydrogen bonding between phenol and aniline in the adsorbed layer would reduce the affinity of each for the aqueous phase and thus increase the tendency of the complex to be expelled from the bulk phase to the interface.[47] Alternatively, a primary adsorbed layer of one solute may hold a further layer of the second solute.[45] The necessary strong attractive forces between the molecules would be due to hydrogen-bonding which, for steric reasons, might favour the second alternative in a close-packed film.

A more recent treatment of the surface tension of ternary solutions is given by Semenchenko.[48]

THE SURFACES OF LIQUID METALS

Although most studies of the surface of liquid mixtures have been made at temperatures close to room temperature, a number of measurements of the surface tensions of liquid metal alloys have been made recently at much higher temperatures.[49] These could be made the basis of adsorption studies by applying the principles outlined in this chapter. Experimentally, the sessile drop method has been used.[50] Results for single metals have shown that the surface tension is very susceptible to the presence of small traces of impurity (as is well known with mercury at room temperature); examples are shown in Fig. 11.13. It can be calculated[51] that a complete monolayer of oxygen is formed on the surface of iron at a concentration in the metal of 0.04%, and for sulphur at a concentration of less than 0.02%.

Results obtained up to 1957 are summarized by Semenchenko.[52] Examples of the work which has been carried out subsequently are given in papers by Addison[53] and Pokrovskii.[54]

The surface tensions have very high values, e.g. 1835, 1924, and 1936 dynes/cm. for iron, nickel, and cobalt, respectively,[55] at 1550°C. Lower values (450–550 dynes/cm.) are found for alloys of lead with tin and with indium[56] in the region 350–550°C. A model has been proposed in which it is assumed that the surface layer is one atom thick. It appears to fit the experimental data satisfactorily.

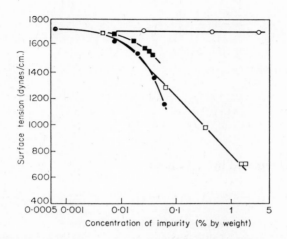

FIG. 11.13.   Effect of impurities on the surface tension of iron[51] at 1570°C: ○, carbon; ■, nitrogen; ●, oxygen; □, sulphur. (Reproduced with the permission of the American Chemical Society from the *Journal of Physical Chemistry*.)

A JUSTIFICATION OF THE FREUNDLICH EQUATION

The Freundlich equation can be derived by combining an expression for the free energy of the surface with the Gibbs adsorption equation.[57] Thus if σ_0 ergs/sq. cm. is the free surface energy of the surface in contact with the pure solvent, and σ_1 that of the surface covered with a monolayer of solute, then σ, the free surface energy of the surface when a fraction θ is covered with solute is given by:

$$\sigma = \sigma_0(1-\theta)+\sigma_1\theta = \sigma_0-(\sigma_0-\sigma_1)\theta. \tag{11.34}$$

Now

$$\theta = \left(\frac{x}{m}\right)\bigg/\left(\frac{x}{m}\right)_m,$$

where $(x/m)_m$ is the corresponding monolayer capacity.
Hence:

$$\sigma = \sigma_0 - \frac{(\sigma_0-\sigma_1)\dfrac{x}{m}}{\left(\dfrac{x}{m}\right)_m}. \tag{11.35}$$

For dilute solutions, $\dfrac{x}{m}$ can be equated with the Gibbs surface excess, Γ, whence:

$$\frac{x}{m} = -\frac{c}{RT} \cdot \frac{d\sigma}{dc} = \frac{c}{RT} \cdot \frac{\sigma_0 - \sigma_1}{\left(\dfrac{x}{m}\right)_m} \cdot \frac{d\left(\dfrac{x}{m}\right)}{dc},$$

and on integration:

$$\ln\left(\frac{x}{m}\right) = \frac{RT\left(\dfrac{x}{m}\right)_m}{\sigma_0 - \sigma_1} \ln c + \ln k. \tag{11.36}$$

If $\dfrac{RT\left(\dfrac{x}{m}\right)_m}{\sigma_0 - \sigma_1}$ is replaced by $\dfrac{1}{n}$, we have

$$\frac{x}{m} = kc^{1/n},$$

which is the usual form of the Freundlich equation.

An essential feature of this derivation is that it is based on an approximation which is only applicable to dilute solutions, viz. that $\dfrac{x}{m} = \Gamma$. This limitation of the Freundlich equation has generally been overlooked. The derivation has been included in this chapter because it is based on the Gibbs equation and is applicable to interfaces generally, subject to the above limitation.

REFERENCES

1. J. H. Hildebrand and R. L. Scott, "The Solubility of Nonelectrolytes", Reinhold Publishing Corporation, New York, 1950, 3rd edition.
2. E. A. Guggenheim and N. K. Adam, *Proc. Roy. Soc.*, 1933, A **139**, 218.
3. A. W. Adamson, "Physical Chemistry of Surfaces", Interscience, New York and London, 1960.
4. W. F. K. Wynne-Jones, *Phil. Mag.*, 1931, [7], **12**, 907.
5. R. B. Dean, *J. Phys. Chem.*, 1951, **55**, 611.
6. J. J. Kipling, "Proceedings of the Third International Congress of Surface Activity", 1960, Verlag der Universitätsdruckerei, Mainz, Vol. 2, p. 77.
7. H. H. King and R. W. Wampler, *J. Amer. Chem. Soc.*, 1922, **44**, 1894.
8. F. E. Bartell and F. C. Benner, *J. Phys. Chem.*, 1942, **46**, 847.
9. J. H. Mathews and A. J. Stamm, *J. Amer. Chem. Soc.*, 1924, **46**, 1071, 2880.
10. P. V. Cornford, J. J. Kipling, and E. H. M. Wright, *Trans. Faraday Soc.*, 1962, **58**, 74.
11. J. A. V. Butler and A. Wightman, *J. Chem. Soc.*, 1932, 2089.
12. R. K. Schofield and E. K. Rideal, *Proc. Roy. Soc.*, 1925, A **109**, 57.
13. J. C. Eriksson, *Adv. Chem. Phys.*, 1964, **6**, 145.

14. A. H. Ellison and W. A. Zisman, *J. Phys. Chem.*, 1959, **63**, 1121.
15. J. W. Breitenbach and H. Edelhauser, *Monatsh.*, 1953, **84**, 384.
16. D. C. Jones and L. Saunders, *J. Chem. Soc.*, 1951, 2944.
17. W. D. Harkins and E. H. Grafton, *J. Amer. Chem. Soc.*, 1925, **47**, 1329.
18. W. D. Harkins and R. W. Wampler, *J. Amer. Chem. Soc.*, 1931, **53**, 850.
19. J. A. V. Butler, *Proc. Roy. Soc.*, 1932, A **135**, 348.
20. J. J. Kipling, *J. Colloid Sci.*, 1963, **18**, 502.
21. A. B. Taubman, *Zhur. fiz. Khim.*, 1952, **26**, 389, cf. *Chem. Abs.*, 1953, **47**, 4687.
22. A. B. Taubman and S. I. Burshtein, *Colloid J.* (*U.S.S.R.*), 1958, **20**, 503.
23. F. M. Fowkes and W. D. Harkins, *J. Amer. Chem. Soc.*, 1940, **62**, 3377.
24. Y. Fu and F. E. Bartell, *J. Phys. Chem.*, 1950, **54**, 537.
25. S. Brunauer, "Physical Adsorption of Gases and Vapours", Oxford University Press, London, 1944.
26. D. G. Dervichian, *Kolloid Z.*, 1956, **146**, 96.
27. A. Couper and D. D. Eley, *J. Polymer Sci.*, 1948, **3**, 345.
28. A. Katchalsky and I. Miller, *J. Phys. Chem.*, 1951, **55**, 1182.
29. J. T. Davies, *J. Colloid Sci.*, 1954, **9**, S 1, 9.
30. T. Kawai, *J. Polymer Sci.*, 1959, **35**, 401.
31. H. L. Frisch and S. Al-Madfai, *J. Amer. Chem. Soc.*, 1958, **80**, 3561.
32. S. Al-Madfai and H. L. Frisch, *J. Amer. Chem. Soc.*, 1958, **80**, 5613.
33. H. L. Frisch and R. Simha, *J. Chem. Phys.*, 1957, **27**, 702.
34. J. Traube, *Annalen*, 1891, **265**, 27.
35. J. Guastalla, *J. Colloid Sci.*, 1956, **11**, 623.
36. E. Duclaux, *Ann. Chem. Phys.*, 1878, [5], **13**, 76.
37. I. Langmuir, *J. Amer. Chem. Soc.*, 1917, **39**, 1848.
38. B. von Szyszkowski, *Z. phys. Chem.*, 1908, **64**, 385.
39. R. H. Aranow and L. Witten, *J. Chem. Phys.*, 1958, **28**, 405.
40. M. A. Higgs, *J. Chem. Phys.*, 1961, **35**, 1504.
41. R. H. Aranow and L. Witten, *J. Chem. Phys.*, 1961, **35**, 1504.
42. A. F. H. Ward, *Trans. Faraday Soc.*, 1946, **42**, 399.
43. L. Guastalla, *Ann. Physique*, 1960, 131.
44. B. Tamamushi, *Bull. Chem. Soc. Japan*, 1932, **7**, 168.
45. J. A. V. Butler and C. Ockrent, *J. Phys. Chem.*, 1930, **34**, 2841.
46. C. Wagner, *Z. phys. Chem.*, 1929, A **143**, 389.
47. C. Ockrent and J. A. V. Butler, *J. Phys. Chem.*, 1930, **34**, 2297.
48. V. K. Semenchenko and V. B. Lazarev, *Bull. Acad. Sci., U.S.S.R.*, 1962, 1997.
49. W. D. Kingery, "Property Measurements at High Temperatures", Wiley, New York, 1959, Chapter 14.
50. W. D. Kingery and M. Humenick, *J. Phys. Chem.*, 1953, **57**, 359.
51. F. A. Halden and W. D. Kingery, *J. Phys. Chem.*, 1955, **59**, 557.
52. V. K. Semenchenko, "Surface Phenomena in Metals and Alloys" (translated by N. G. Anderson), Pergamon Press, Oxford, 1961; original Russian edition, Moscow, 1957.
53. C. C. Addison, *J. Chem. Soc.*, 1962, 3217.
54. N. L. Pokrovskii and D. S. Tissen, *Proc. Acad. Sci.* (*U.S.S.R.*), 1959, **128**, 1228.
55. P. Kozakevitch and G. Urbain, *Compt. rend.*, 1957, **244**, 335; *J. Iron Steel Inst.*, 1957, **186**, 167.
56. T. P. Hoar and D. A. Melford, *Trans. Faraday Soc.*, 1957, **53**, 315.
57. D. C. Henry, *Phil. Mag.*, 1922, [6], **44**, 689.

CHAPTER 12

Adsorption at the Liquid–Liquid Interface

TYPES OF INTERFACIAL LAYER

Three main types of adsorption have been studied at this interface. Films of substances insoluble in both liquids will not be considered here. They are analogous to the insoluble films which can be formed at the liquid–vapour interface. Although of great interest and importance, they are not strictly examples of adsorption from solution.

The second type is adsorption of a substance which is soluble in one liquid but insoluble in the other. In some ways this is analogous to adsorption at the liquid–solid interface. The third type is adsorption of a substance soluble in both liquids. The presence of a third component at this interface is of major importance in the stabilizing of emulsions, about which a great deal is known generally. The process of adsorption as such, however, has not been considered extensively from the standpoint adopted in this book.

The liquid–liquid interface appears in some ways to be simpler than the liquid–solid interface, although there are some experimental difficulties in investigating it.[1] If pure liquids are used, the interface is homogeneous and there is no uncertainty about the surface area. There might be considerable uncertainty, however, about the specific orientation which molecules of each liquid adopt at the interface. Orientation at each side of the interface must be expected, giving this region "a degree of organized structure" not shown by either bulk liquid phase.[2]

THIRD COMPONENT SOLUBLE IN ONE LIQUID

Water is of outstanding interest as one liquid phase, particularly because it has a very high interfacial tension against non-polar liquids such as hydrocarbons, but its properties even as a pure liquid are complicated and are incompletely understood.

The complexity of interfacial water is shown by the addition of the lower aliphatic alcohols to the water–hexane interface. Figure 12.1 shows the effect of ethanol; methanol and propanol produce similar effects. At low concentrations there is a sharp increase in interfacial tension at 25°c, with a fall at higher concentrations. No such increase, however, occurs at 40°c with the three lower alcohols or with butanol at either temperature. Over

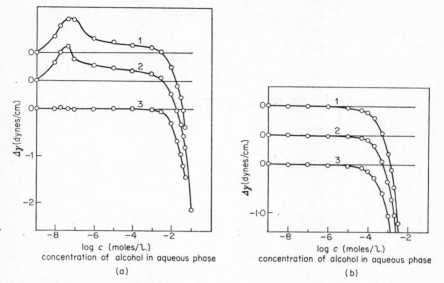

Fɪɢ. 12.1. Effect of a third component on the interfacial tension between water and n-hexane[2]: (a) ethanol at (1) 10°c, (2) 25°c, (3) 40°c; (b) n-butanol at (1) 10°c, (2) 25°c, (3) 40°c. (Reproduced with permission from the *Journal of the Chemical Society*.)

this range of temperature, a change occurs in the structure of water, which is reflected in a sharp fall in its interfacial tension against hexane in the absence of solutes. The abnormal increase caused by the alcohols might be due to their repulsion from a "rigid" interfacial water layer, into which they can break at higher concentrations. At higher temperatures this layer loses its rigidity and the alcohols have the more normal effect. A similar effect has been noticed in the behaviour of aqueous solutions at mercury electrodes. A "quasi-crystalline" region forms at the interface at low temperatures and under anodic conditions, which inhibits the entry of ions of normal energy. At higher temperatures or under cathodic conditions the effect is no longer observed.[3]

Hutchinson used the exact form of the Gibbs equation to calculate the composition of the interface between water and solutions of fatty acids and of fatty alcohols in benzene.[4] The results were interpreted in terms of a unimolecular layer, with the aliphatic molecules oriented essentially perpendicular to the interface. Application of the method developed by Guggenheim and Adam for the liquid–vapour interface (see Chapter 11) gives the total concentration of the acid or alcohol in the interface; the area per molecule can then be calculated. For the alcohols (n-hexanol to n-decanol) the areas are a little more than 20 sq. Å/molecule at the highest concentration studied. Thus the alcohols tend, with increasing concentration, to the formation of a complete monolayer with the molecules oriented perpendicular to the surface. The same conclusions can be drawn from a further study on n-octanol and n-decanol,[5] although there are slight discrepancies between the two sets of experimental results.

By contrast, the areas occupied by the fatty acids (n-butyric to lauric) are much higher. Figure 12.2 shows the much lower surface excesses of the

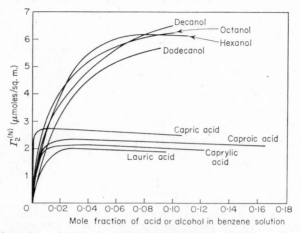

FIG. 12.2. Adsorption of aliphatic acids and alcohols at the water–benzene interface at 23°c, plotted from Hutchinson's data.[4]

acids, a result which originally suggested a much less highly condensed packing.[4] The adsorption of each acid, however, clearly approaches a limiting value at high concentrations, when adsorption is studied from the aqueous layer. Adsorption from heptane, at the interface with water, gives a similar though not identical limiting value for caproic acid.[6] The isotherm in Fig. 12.3 can probably be extrapolated to zero concentration (cf.

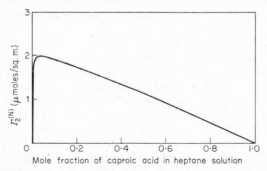

FIG. 12.3. Adsorption of caproic acid at the water–heptane interface[6] at 20°c. (Reproduced from *Monatshefte für Chemie*.)

Chapter 7) to give the monolayer value; this would give a molecular area of about 41 sq. Å/monomeric molecule. Both of these results seem to be most satisfactorily explained on the assumption that the molecules lie parallel to the interface in dimeric form.† The difference between the acids and alcohols could be attributed to the double hydrogen-bonding which occurs in the dimeric acid molecules, reducing their polar character to a much greater extent than can occur with alcohol molecules.

Support for this suggestion comes from the observation[7] that fatty acids are more surface active at the water–benzene interface at pH 7·6 than at pH 6·5. The ions (present in higher proportion at pH 7·6) are more likely than the neutral molecules to be oriented perpendicular to the interface. The parallel orientation would also explain the further observation[8] that the limiting adsorption of caprylic acid is the same at 10°c and 40°c. This would be unlikely unless in each case a complete monolayer were being formed. (The same effect is found for n-octyl alcohol at the two temperatures, in accordance with the suggestion that it forms a complete monolayer with the perpendicular orientation.)

The effect of the surface structure of the solvent is shown by the adsorption of n-caprylic acid from tetradecane which is much greater than its adsorption from benzene.[8] The extent is only compatible with the formation of a monolayer if the perpendicular adsorption is adopted. In this case it might be supposed that the aliphatic chains of the solute and solvent would lie together, the attraction of the carboxyl group for the water determining the orientation of both. The formation of stable oriented, mixed monolayers containing stearic acid and n-hexadecane at water surfaces[9] supports this suggestion. Such an alignment of solute and solvent could not be as extensive with benzene; any orientation of the benzene might, indeed, be to present its π-electron cloud parallel to the water surface, so encouraging a parallel orientation of the fatty acid. The carboxyl group of the acid could then be satisfied by the formation of dimeric molecules, the normal form in benzene solutions. By contrast, hydrogen-bonding in the alcohols does not lead specifically to dimeric molecules, nor would the dimeric alcohol molecules be so suited geometrically to the parallel orientation. In this case, therefore, formation of the dimeric molecule presents less advantage as compared with hydrogen-bonding to the water surface.

The effect of solvation of long-chain polar compounds can also be seen in the minimum molecular area occupied at the interface. At the octane–water interface, for example, stearic acid occupies a minimum of 30·7 sq. Å/molecule,[10] compared with 20·5 sq. Å/molecule in compressed insoluble films on water. The nature of the solvent may be important not only in this respect, but also in the extent to which it can be polarized at the interface with water. Thus adsorption of long-chain compounds is much weaker from benzene than from cyclohexane at the interface with water (Fig. 12.4).[10]

† Dr. Hutchinson (private communication) agrees that this is a possible orientation.

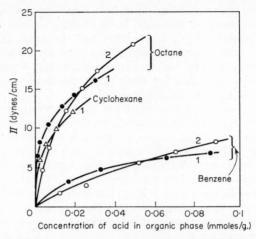

FIG. 12.4. Surface activity of: (1) stearic acid, (2) lauryl alcohol, at interfaces between water and benzene, cyclohexane, and octane.[10] [Reproduced from *Colloid Journal* (*U.S.S.R.*)]

The effect of the polar group of the solute can be seen in adsorption of phenol derivatives at the hexane–water interface.[10] Hydroxyphenols show increasing surface activity with increasing symmetry ($o < m < p$), but o-nitrophenol is much less active than either m- or p-nitrophenol. This is attributed to internal hydrogen bonding in o-nitrophenol, which reduces the polarity and hence the strength with which it is adsorbed at the water surface.

A few results have been obtained for the adsorption of polymers at water–benzene interfaces. They suggest that the adsorbed molecules have a greater degree of flexibility at this interface than at the water–air interface.[11, 12]

Equations of State

From values of the surface excess, Γ, the areas occupied by molecules of the third component can be obtained, and related to the surface pressure, $\pi(= \gamma_0 - \gamma)$, exercised by the film. Thus a comparison can be made between such films and those at the liquid–vapour interface and with insoluble films on aqueous substrates. There is some uncertainty about the accuracy of such results, but at least some qualitative conclusions can be drawn. Films of fatty acids and alcohols at the benzene–water interface are condensed, and a phase change is apparent in some curves[1, 5] (Fig. 12.5). For the regions of lower pressures the Langmuir equation,

$$(\pi - \pi_0)(A - A_0) = C \tag{12.1}$$

is obeyed, and gives satisfactory values for A_0, the molecular area of the component forming the film. The inter-molecular attraction, as given by π_0, is small in this region.

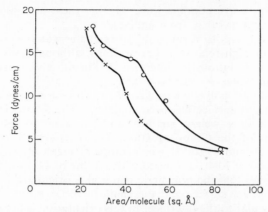

FIG. 12.5. Force–area curves, showing a phase change, for alcohols adsorbed at the water–benzene interface: ○, octanol; ×, decanol.[5] (Reproduced from the *Journal of Colloid Science.*)

Effect of Temperature

The interfacial tension between toluene and water decreases linearly with rise in temperature. When the toluene contains dissolved hexachlorethane, however, a discontinuity is observed[13] at temperatures in the region of 43°C, which is the temperature of transition from the orthorhombic to the triclinic form of solid hexachlorethane. The area occupied in the adsorbed film at low temperatures corresponds to the size of the molecule in the orthorhombic crystal, and that at high temperatures to the size of a molecule capable of free rotation. The difference in energy in the two adsorbed films also corresponds to the energy of free rotation.

Adsorption at Mercury Surfaces

The water–mercury interface provides a special case of adsorption of components soluble in one phase only. It has been extensively investigated in polarographic studies involving the adsorption of ions. In a few cases, the results are relevant to this chapter because the adsorbates have been weak electrolytes. In non-polarographic work, the adsorption of non-electrolytes at this interface has also been examined.

Adsorption of polar and of aromatic compounds (e.g. aniline, benzoic acid) from aqueous solution is very much stronger at the solution–mercury interface than at the solution–vapour interface. From comparisons of the concentrations needed to give the same lowering of surface tension at the two interfaces, standard free energy changes of 1–2 kcal./mole for the transition between the two interfaces have been calculated.[14] Aromatic compounds are adsorbed more strongly than the corresponding saturated ring compounds. The neutral molecules are also much more strongly adsorbed than the corresponding ions, whether these are anions (e.g. from benzoic acid) or cations (e.g. from aniline).

Caproic acid appears to form multilayers at the water–mercury interface from solutions of high relative concentrations, although only a monolayer is formed at the solution–air interface.[15] Phenol also appears to form multilayers at the water–mercury interface.

The adsorption of polyethylene glycol of various molecular weights at the water–mercury interface has been studied.[16] The interfacial tension was measured by the drop-weight method, and was used to calculate surface excess by means of the Gibbs equation. The area per molecule adsorbed varied between 36 and 49 sq. Å over a range of mean molecular weight of approximately 300 to 7000, compared with a value of 44·3 sq. Å for ethylene glycol. This implies that the polymeric material is oriented with the terminal group at the mercury surface and the chain in the aqueous phase approximately perpendicular to the interface. In this orientation it occupies a greater area than the hydrocarbon chain in, for example, close-packed films of fatty acids. This is attributed to the hydration of the ether groups in the chain, which is also responsible for the solubility of the polymers in the aqueous phase. The orientation differs from that at the water–air interface (Chapter 11).

Interfacial tensions have also been used to examine adsorption at this interface. The composition of the interfacial film is given by the equation:

$$\gamma_{12} = n_1^s \gamma_1 + n_2^s \gamma_2, \tag{12.2}$$

where γ_{12} is the interfacial tension of a binary liquid in contact with mercury, γ_1 and γ_2 being, respectively, the interfacial tensions due to the two pure components; n_1^s and n_2^s are the mole fractions of the respective components in the interfacial film.[17] On this basis, it is seen that the aryl halides are adsorbed preferentially to cyclohexane at the mercury surface.

A few data are available on the adsorption of molecules of biological origin. Denatured deoxyribonucleic acid (DNA) and ribonucleic acid (RNA) are adsorbed at the dropping mercury electrode with almost identical results.[18] From a solution of pH 5·5 (with solute of molecular weight $7·9 \times 10^5$) DNA was adsorbed with a limiting area of 95 sq. Å per nucleotide unit (of molecular weight 330). This is $2\frac{1}{2}$ times larger than the area which would be required if the double helix structure were maintained, and suggests that the DNA when adsorbed is unfolded.

THIRD COMPONENT SOLUBLE IN BOTH LIQUIDS

An interesting investigation has been made of the adsorption of short-chain polar molecules at the interface between water and petroleum ether.[19] The data were fitted (Fig. 12.6) by means of an equation of state derived by Langmuir:

$$\pi (A - A_0) = kT, \tag{12.3}$$

where π is the surface pressure, A is the area occupied by each molecule at that surface pressure, and A_0 is the limiting area per molecule in the interface. The limiting area for the lower aliphatic alcohols was 18·5 sq. Å/molecule in close agreement with the crystallographic value of 18·4 sq. Å for the cross-sectional area of the hydrocarbon chain. This implies that the alcohol

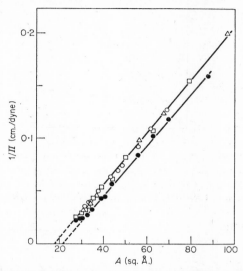

FIG. 12.6. Verification of the equation of state for adsorption at the water–petroleum ether interface at 20°c: ●, n-butyric acid; ○, n-butyl alcohol; □, n-propyl alcohol; △, ethyl alcohol.[19] (Reproduced with the permission of the Royal Society from *Philosophical Transactions*.)

molecules are oriented with the hydrocarbon chain perpendicular to the interface. The corresponding limiting area for butyric acid was 21·5 sq. Å/molecule.

The adsorption isotherm corresponding to equation (12.3) is:

$$\frac{A_0}{A-A_0}\, e^{A_0/(A-A_0)} = a\, e^{-\Delta G/kT}, \tag{12.4}$$

where a is the activity of the solute in the aqueous phase and ΔG the free energy of adsorption of the molecule at the interface. The corresponding form of the Langmuir adsorption equation is:

$$\frac{A_0}{A-A_0} = a\, e^{-\Delta G/RT}. \tag{12.5}$$

Equation (12.5) is appropriate to an immobile adsorbed film, whereas equation (12.4) is appropriate to a mobile film. In the above experiments it was found that equation (12.4) was obeyed (Fig. 12.7). The data are plotted to give a straight line for each equation if it is obeyed.

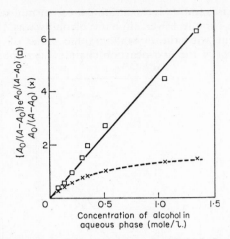

FIG. 12.7. Isotherms (see text) for adsorption of n-propyl alcohol at the water–petroleum ether interface: □, for a mobile film; ×, for a non-mobile film.[19] (Reproduced with the permission of the Royal Society from *Philosophical Transactions*.)

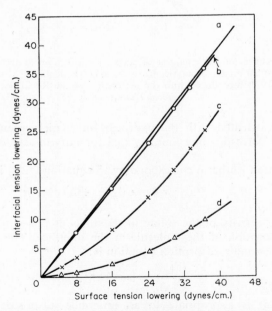

FIG. 12.8. Effect of ethyl alcohol in lowering interfacial tension at: (a) water–air, (b) water–heptane, (c) water–benzene, (d) water–methyl n-amyl ketone interfaces.[21] (Reproduced with the permission of the American Chemical Society from the *Journal of Physical Chemistry*.)

In older work, Harkins and King used the Gibbs equation to calculate the adsorption of butyric acid at the benzene–water interface.[20] They claimed that, at low concentrations in the aqueous phase, there was much less adsorption at the water–benzene interface than at the water–air interface. At higher concentrations, a complete monolayer was formed at both interfaces, with almost identical values for the area occupied by the butyric acid molecule (36·1 and 35·9 sq. Å, respectively). These figures are so high

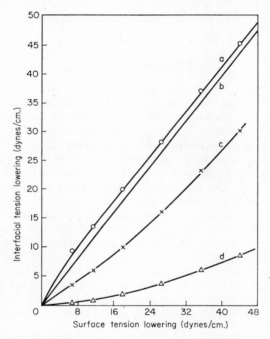

Fig. 12.9. Effect of n-butyl alcohol in lowering interfacial tension at: (a) water–air, (b) water–heptane, (c) water–benzene, (d) water–methyl n-amyl ketone interfaces.[21] (Reproduced with the permission of the American Chemical Society from the *Journal of Physical Chemistry*.)

that it would be desirable to re-calculate them. Unfortunately the full data required do not appear to be given in the original paper.

Bartell claims that adsorption of butyric acid is almost the same at the water–paraffin interface as at the water–air interface, and that similar results are obtained for the lower fatty alcohols.[21] These conclusions are based on a comparison of the lowering of interfacial tensions and surface tensions, respectively. Lower adsorption occurs when the second liquid has a lower interfacial tension against water; compare benzene (34·3 dynes/ cm.) with methyl n-amyl ketone (12·5 dynes/cm.), Figs. 12.8 and 12.9.

An explanation of the similar effects at the water–heptane (50·9 dynes/cm.) and water–air (72·0 dynes/cm.) interfaces is given in terms of the free energy change. If a solute displaces water completely at the interface with air, a free surface energy of "at least 22 ergs/sq. cm." remains. Thus the adsorption process is accompanied by a change of only about 50 ergs/sq. cm. At the water–heptane interface, however, complete displacement of the water by solutes of the kind discussed would be accompanied by a change in free energy corresponding to almost the whole of the 50·9 ergs/sq. cm. On this basis the two interfaces are very similar in respect of their ability to effect adsorption.

TRAUBE'S RULE

The qualitative applicability of Traube's Rule (Chapter 11) to this interface can be deduced from Harkins' early results on the adsorption of the lower fatty acids at the water–benzene interface.[22] Adsorption increases from formic acid to butyric acid at low concentrations. The Traube coefficient lies between 3 and 4 for propionic and higher acids.[23]

The quantitative application is shown in Haydon and Taylor's data for adsorption of the fatty alcohols from aqueous solution at the water–petroleum ether interface. The free energy of adsorption increases linearly with increasing chain-length, corresponding to an increase of -820 cal./mole for each CH_2 group. This may be compared with a value of -625 cal./mole for the air–water interface. By contrast, differences in free energy of adsorption of lauric, palmitic, and stearic acids from hexane at the hexane–water interface are almost negligible.[24] In this case the adsorption process involves a change of environment of the carboxyl group (from hexane to water), but not of the hydrocarbon chain. Similarly, the energy of adsorption of the long-chain alcohols at the octane–water interface is only 50–100 cal./mole for each CH_2 group.[10] This aspect of Traube's Rule has also been considered by Vignes.[25]

Multicomponent Systems

Competitive adsorption of charged species at the water–mercury interface is important in electrochemistry, but less is known of the adsorption of uncharged species. The general situation can be deduced from the principle that the surface energy at any interface tends to a minimum. Suppose that A causes a greater lowering of the surface tension of a solvent than B. Then in a solution containing both A and B, it is likely that B will not be adsorbed, for if it were to be adsorbed in competition with A, the maximum lowering of surface tension would not occur.[26]

For two solutes of approximately equal surface activity, the lowering of surface tension tends to be greater than that of either solute separately, but less than their sum.

The adsorption of each of stearic acid and urea at the water–hydrocarbon interface is enhanced in the presence of the other.[27] This is attributed to specific interaction between the polar groups, analogous to that found between those of phenol and aniline at the water–air interface (Chapter 11).

REFERENCES

1. E. Hutchinson and D. Randall, *J. Colloid Sci.*, 1952, **7**, 151.
2. F. Franks and D. J. G. Ives, *J. Chem. Soc.*, 1960, 741.
3. D. C. Grahame, *J. Chem. Phys.*, 1955, **23**, 1725.
4. E. Hutchinson, *J. Colloid Sci.*, 1948, **3**, 219.
5. N. Pilpel, *J. Colloid Sci.*, 1956, **11**, 51.
6. F. Seelich, *Monatsh.*, 1948, **79**, 348.
7. J. G. Dimenstein, *Anales Fac. Qúim. Farm. Univ. Chile*, 1956, **8**, 127.
8. E. Hutchinson, *J. Colloid Sci.*, 1948, **3**, 235.
9. H. D. Cook and H. E. Ries, *J. Phys. Chem.*, 1959, **63**, 226.
10. A. B. Taubman and S. I. Burshtein, *Colloid J. (U.S.S.R.)*, 1958, **20**, 503.
11. J. T. Davies, *J. Colloid Sci.*, 1954, **9**, S 1, 9.
12. T. Kawai, *J. Polymer Sci.*, 1959, **35**, 401.
13. E. Ferroni and G. Gabrielli, *Atti Accad. naz. Lincei, Rend. Classe Sci. fis. mat. nat.*, 1963, [8], **34**, 161.
14. A. N. Frumkin, R. I. Kaganovich, and E. S. Bit-Popova, *Proc. Acad. Sci. (U.S.S.R.)*, 1961, **141**, 899.
15. A. Frumkin, A. Gorodetzkaja, and P. Tschugunoff, *Acta Physicochim. U.R.S.S.*, 1934, **1**, 12.
16. A. Couper and D. J. Priest, "Third International Congress of Surface Activity", Verlag der Universitätsdruckerei, Mainz, 1960, Vol. II, p. 77.
17. K. L. Wolf and A. W. Neumann, *Z. phys. Chem. (Leipzig)*, 1962, **219**, 60.
18. I. R. Miller, "Proceedings of the Third International Congress of Surface Activity", Verlag der Universitätsdruckerei, Mainz, 1960, Vol. II, p. 92.
19. D. A. Haydon and F. H. Taylor, *Phil. Trans.*, 1960, **252**, 225.
20. W. D. Harkins and H. H. King, *J. Amer. Chem. Soc.*, 1919, **41**, 970.
21. F. E. Bartell and J. K. Davis, *J. Phys. Chem.*, 1941, **45**, 1321.
22. W. D. Harkins and E. C. Humphery, *J. Amer. Chem. Soc.*, 1916, **38**, 242.
23. S. Boas-Traube and M. Volmer, *Z. phys. Chem.*, 1937, A **178**, 323.
24. A. F. H. Ward and L. Tordai, *Rec. Trav. chim.*, 1952, **71**, 482.
25. A. Vignes, *J. Chim. phys.*, 1960, 966.
26. C. Ockrent and J. A. V. Butler, *J. Phys. Chem.*, 1930, **34**, 2297.
27. A. B. Taubman, *Doklady Akad. Nauk S.S.S.R.*, 1950, **74**, 521, cf. *Chem. Abs.*, 1951, **45**, 417.

CHAPTER 13

Kinetics of Adsorption from the Liquid Phase

INTRODUCTION

As in the field of gaseous adsorption, much less attention has been paid to the kinetics of the adsorption process than to the state of equilibrium ultimately attained. Indeed, most of the available information about the rate of adsorption from solution seems to have been obtained primarily to determine what time of contact must be allowed for attainment of equilibrium. This type of study, although limited, has produced some interesting results.

The main features of this chapter are discussed in terms of the liquid–solid interface. Important aspects of the other two interfaces are considered in the later sections.

Time Taken to Reach Equilibrium

Adsorption of simple molecules onto a plane solid surface normally takes place rapidly if the solution is not viscous. In adsorption from aqueous solutions of butanol onto Graphon, equilibrium was established within 10 minutes.[1] For larger molecules a longer time is required. In adsorption of stearic acid from benzene by iron powder, equilibrium was established within 4 hours, but 90% of the total adsorbate had reached the surface within 5 minutes.[2]

Adsorption by porous solids may also be fast in some cases, provided that the liquid is agitated continuously. Thus for a silica gel, equilibrium with aqueous solutions of mono-, di-, and tri-chloracetic acid was established within $1\frac{1}{2}$ to $2\frac{1}{2}$ minutes with continuous agitation, but 12 to 30 minutes was required when agitation was only intermittent. Much longer times were required with a charcoal,[3] which suggests that the time taken to reach equilibrium depends on the pore size of the adsorbent in relation to the molecular size of the adsorbate.

The adsorption of high polymers is a little slower. The amount of GR–S rubber adsorbed by Graphon from benzene remained essentially constant after 1 hour, but displacement of the lower molecular weight by the higher molecular weight fractions continued up to about 48 hours.[4] Adsorption of rubbers by commercial carbon blacks, however, may not reach equilibrium for 100 hours.[5] This type of adsorption may be of importance in biological systems. In a pilot study related to the coagulation of blood, the adsorption of dextran by collodion was found to be incomplete after 50 hours. The time required for equilibrium to be established and the extent of adsorption were affected by the presence of added substances such as salts and detergents.[6]

Studies of the rate of adsorption of polymers, however, mean little unless the adsorbent has been well characterized; this is particularly so with porous adsorbents. Thus adsorption of polystyrene from methyl ethyl ketone containing 1 % of methanol is described as being complete in about 40 minutes with one particular charcoal.[7] The time is only significant when the nature of the charcoal is considered; in this case it was decolorizing charcoal and therefore probably had relatively wide pores. A much longer time would probably be required for a charcoal of the type used in gas adsorption and having considerably narrower pores. For the same system, the absolute rate of adsorption changed with the nature of the solvent, becoming much greater as the concentration of methanol in the methyl ethyl ketone increased, i.e. as the solvent became a progressively poorer solvent for the polymer. Adsorption from the poorest solvents might have involved adsorption of aggregates rather than individual polymer molecules.

The rate may also depend on the molecular weight of the polymer. The time required for "complete" adsorption of polyvinyl acetate by charcoal increased as the molecular weight decreased[8] from 170,000 to 22,000. This may reflect the possibility of diffusion at a perceptible rate into the narrower pores for the smaller molecules but not for the larger molecules. In accordance with this, it is found that in each case the rate of adsorption is initially constant for each fraction (a different rate being observed for each). For the highest molecular weight fraction it remains constant for almost the whole course of adsorption, whereas for that of the lowest molecular weight, the second half of the material adsorbed is taken up very much more slowly than the first half.

Adsorption onto a plane surface from dilute polymer solution appears to be governed by a kinetic law similar to that postulated by Langmuir for adsorption of gases.[9] In some systems, however, adsorption can occur in two stages (Fig. 13.1), each stage being governed by the same general equation (see below). This was found in the adsorption of polystyrene from methyl ethyl ketone by charcoal,[7] though only when the solvent contained a small quantity of water.[10] Methyl alcohol had an effect similar to that of water, but higher concentrations were needed. The break, or "resting period" was attributed to reorientation of the molecular layer first adsorbed, so leaving further surface of the adsorbent free, or to migration of the layer initially adsorbed on the exterior of the charcoal into the pores. The duration of the first stage increased as the molecular weight of the polymer increased; this would be in accordance with both mechanisms.

Exchange Processes

A few measurements have been made of the rate of exchange between adsorbed molecules and those of the same species present in the solution at

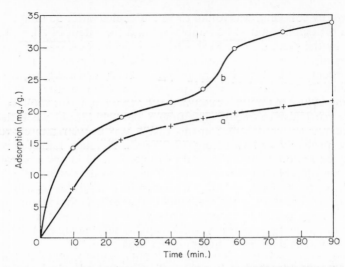

FIG. 13.1. Adsorption of polystyrene by charcoal from methyl ethyl ketone as a function of time[10]: (a) from pure solvent, (b) in presence of water (1% v/v). (Reproduced from the *Journal of Polymer Science*.)

equilibrium. Such experiments are readily carried out if the molecular species in one phase is made radioactive.

The rate of exchange between benzoic acid adsorbed on a carbon surface and benzoic acid in aqueous solutions depends very markedly on the accessibility of the adsorbing surface.[11] Thus for the non-porous Graphon, exchange was complete in about two minutes, and for highly activated Saran charcoals was almost complete within an hour. For a typical gas-mask charcoal, however, two days were needed, and for an unactivated Saran carbon, only 40% exchange had occurred after eight days. The authors attributed these results partly to the slow rate of diffusion which occurs in very narrow pores (such as are known to be present in unactivated but porous carbons) and to the strength with which adsorbates are held in narrow pores. The latter effect may be an immediate result of the enhanced field produced by two surfaces in close proximity. Further, the rate at which any molecule can escape from a pore of molecular diameter is very low, as has been emphasized by de Boer.[12] As simple molecules are adsorbed very rapidly at the water–air interface, which is a plane surface, this interpretation of adsorption by porous solids is very likely to be valid. It would be interesting to have more precise measurements of the kinetics of adsorption of simple molecules by non-porous solids.

Extensive exchange of perfluoro-octanoic acid occurs between n-decane solutions and metal or glass surfaces within one minute. Complete exchange is reached after a time which increases with decreasing concentration of the solution. Exchange between this acid and stearic acid under the same conditions is slower,[13] and differs from the situation in which the adsorbed acid is removed as a salt.[14]

RATE CONSTANTS FOR THE OVERALL PROCESS

Swearingen and Dickinson[3] attempted to obtain a rate constant for the process of adsorption. They appear to have assumed that the rate of adsorption was proportional to the distance of the system from equilibrium, and found rate constants which were valid over 75–80% of the total adsorption of acids from aqueous solutions by silica gel and charcoal. As the adsorption process was fast, the rate was followed by measuring the conductivity of the solution.

Similarly an equation (originating with Lagergren[15]):

$$kt = 2 \cdot 303 \log \frac{a_\infty}{a_\infty - a},\qquad(13.1)$$

is followed approximately in the adsorption of polymers by charcoal;[7] a is the amount adsorbed in time t, a_∞ the amount adsorbed at saturation, and k a rate constant.

For the adsorption of p-cresol from aqueous solutions by activated carbon, the equation:

$$a = \frac{a_\infty t}{t + k}\qquad(13.2)$$

is obeyed, where k is the time required for half-saturation of the adsorbent.[16] This can be re-written to give two linear equations:

$$\frac{a}{a_\infty - a} = \frac{t}{k}\qquad(13.3)$$

and

$$\frac{t}{a} = \frac{k}{a_\infty} + \frac{t}{a_\infty}.\qquad(13.4)$$

The distance from equilibrium is important in both adsorption and desorption. For both adsorption and desorption of sodium dodecylbenzene sulphonate by cotton, the following equation applies:

$$-\frac{\mathrm{d}\phi}{\mathrm{d}t} = 2k\phi \sinh b\phi,\qquad(13.5)$$

where ϕ is the distance from equilibrium as a fraction of initial distance from equilibrium. Thus

$$\phi = \frac{a_t - a_\infty}{a_0 - a_\infty},\qquad(13.6)$$

where a_0, a_t, and a_∞ are the amounts adsorbed at time 0, t, and ∞, respectively.[17]

A Langmuir model has been used successfully to describe the adsorption of polyvinyl acetate from benzene onto chrome plate.[9] The equation

$$\frac{d\theta}{dt} = k_1(1-\theta)c - k_{-1}\theta \qquad (13.7)$$

was obeyed for dilute solutions up to surface coverages of more than 50%. In this equation, θ is the fraction of the surface covered by polymer, c is the concentration of polymer in solution, and k_1 and k_{-1} are the rate constants for adsorption and desorption, respectively. The results suggested that the polymer molecules, which existed in solution as coils, were adsorbed in the same form, but then became reoriented to a configuration in which nearly all segments were in contact with the surface; this latter process is believed to be kinetically fast.

Effect of Temperature

Rate constants can be obtained from equation 13.1. The logarithm of the rate constant, if plotted against the reciprocal of absolute temperature, has been found to give a straight line.[10] For adsorption of polystyrene by charcoal from anhydrous methyl ethyl ketone, an activation energy can be calculated which is so small as to indicate that diffusion of the polymer to the surface is the rate-determining step when the system is stirred continuously, i.e. each charcoal granule is surrounded by a diffusion layer.

The Processes of Adsorption

The total process of adsorption can be separated into three stages[18]:

(i) diffusion through the solution to the external surface of the adsorbent, "external" diffusion;

(ii) diffusion within the pores of the adsorbent, "internal" diffusion;

(iii) adsorption at the surface.

In physical adsorption, the third process is so fast that it is never rate-determining. The second stage is only important for porous solids. A mathematical treatment based on these concepts has been put forward,[18–20] but has not yet been examined experimentally.

The Process of Desorption

The use of radioactive labelling has made it possible to study the rate of desorption of stearic acid from a metal plate.[21] The first layer of adsorbed acid (which appears to have been chemisorbed) was not removed by the desorption process adopted. Subsequent layers were removed according to the equation

$$\phi = \phi_E(1 - e^{-kt}), \qquad (13.8)$$

where ϕ is effectively the concentration of the desorbing solution at time t, ϕ_E the concentration at equilibrium, and k a rate constant. It therefore seems likely that desorption, like adsorption, is diffusion-controlled.

THE LIQUID–VAPOUR INTERFACE

When a new surface is created, the surface tension of pure liquids rapidly reaches a steady value (in less than 0·01 sec., and possibly in a much smaller fraction of a second, for water[22] and for mercury[23] at their interfaces with air). At the creation of these interfaces, no more is required than the reorientation of molecules already present. At the surface of a solution, however, the components are usually present in a different proportion from that of the bulk liquid when equilibrium has been established. Transport to and from the surface is therefore required, which is likely to take much longer than molecular reorientation.

Measurements for the liquid–vapour interface are almost entirely confined to aqueous solutions. These include extensive results for ionic surface-active agents, with which this book is only incidentally concerned. In general, two processes may be observed: an initial rapid adsorption, usually lasting

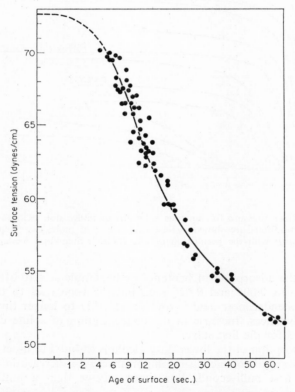

FIG. 13.2. Surface tension of an aqueous solution of n-decyl alcohol (0·012 g./l.) at 20°c, as a function of time.[25] (Reproduced from the *Journal of Colloid Science*.)

for a few seconds or even fractions of a second for small molecules, and a slow aging which may continue for several days or weeks. There is some evidence that the time taken for any stage depends on the apparatus used.[24] Experimental results are expressed sometimes as curves of surface tension against time[25] (Fig. 13.2), and sometimes as curves of surface pressure, π, against time[26] (Fig. 13.3). The surface pressure, or spreading pressure, is the difference between the surface tension of the pure solvent and that of the given solution ($\pi = \gamma_0 - \gamma$).

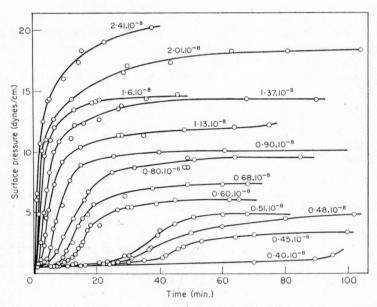

FIG. 13.3. Surface pressure for aqueous solutions of undecanoic acid as a function of time.[26] Concentrations are shown against each curve in moles/c.c. (Reproduced from *Kolloid Zeitschrift* with the permission of Dr. Dietrich Steinkopff Verlag, Darmstadt.)

Most of the adsorption of heptanol or heptanoic acid (0·0188 M) at 20°C occurs between 0·005 and 0·025 sec., but the time rises to thousands of seconds for hendecanoic acid,[27] and presumably to longer times for more complex substances. Increase in the concentration of solute decreases the time required for the first stage.[28]

In the adsorption of the lower fatty alcohols (n-butyl to n-octyl), diffusion control shows signs of increasing importance with increase in chain-length during the first millisecond. For times greater than a millisecond, the equation

$$\ln(\Delta V_E - \Delta V) = -Kt + \text{constant} \tag{13.9}$$

applies, ΔV_E being the surface potential at equilibrium, ΔV that at the time t, and K a constant; the surface potential is proportional to surface concentration. This equation could be taken to imply either diffusion through an unstirred layer at the surface or difficulty in penetrating the existing adsorbed film.[29] The results are consistent both with adsorption according to a Langmuir mechanism and with the application of the Law of Mass Action to the equilibrium between surface and bulk solution, cf. Ward and Tordai.[30]

In a number of experiments, adsorption appeared to be diffusion-controlled, and it has been suggested that the total process can be divided into two stages:

(i) diffusion of molecules from the bulk of the solution into the "sub-surface", i.e. the layer immediately adjacent to the surface layer, followed by

(ii) movement of molecules from the sub-surface into the surface layer, possibly with a particular orientation. For many systems (including aqueous solutions of lower acids and alcohols) the first stage is rate-determining,[31] but for others a major energy barrier appears to limit the rate of penetration into the surface layer. Thus free energies of activation of 13·8, 17·7 and 18·3 kcal./mole have been calculated for the slow stage of adsorption of sebacic, suberic, and azelaic acids, respectively, at the water–air interface.[32]

One possibility is that the first molecules to arrive at the surface are able to assume a random orientation which can be taken up rapidly. Those arriving later, however, can only find a site in a partly-filled surface if those already present adopt a regular instead of a random orientation.[33] This suggestion would presumably be applicable particularly to molecules with chains of substantial length (e.g. sebacic acid in the paper cited). Dervichian considers that there is an energy barrier to adsorption (and thus to desorption) which varies with the amount adsorbed.[26] This conclusion follows from an evaluation of the adsorption and desorption coefficients for the substances which he studied (fatty acids in aqueous solutions), cf. Chapter 11.

The adsorption coefficient is low when the surface concentration is low. This rises suddenly to a very much higher level, at which it remains over the region of concentration at which exchange readily occurs. It then falls suddenly when the molecular area of the adsorbed species has fallen to about 50 sq. Å, at which stage a complete (liquid-like) monolayer is considered to have been formed. The desorption coefficient shows a correspondingly sudden rise, but then falls more gradually as cohesive forces between the adsorbed molecules become increasingly important, finally reaching a low value at the formation of a complete monolayer.

In the adsorption of proteins from aqueous solutions, which is treated as being irreversible, the rate is in accordance with a diffusion equation at low surface pressures, but becomes relatively lower at high surface pressures; an energy barrier arises as the film becomes more closely packed.

Sutherland has pointed out that evidence for diffusion control has come mainly from experiments with vibrating jets and was dependent on the characteristics of the orifice.[35] The rate of change of surface tension would then depend on the hydrodynamics of the system. Hansen and Wallace have suggested minimum specifications to overcome this effect, and conclude that results can be obtained in agreement with those given by other methods.[27] They are, however, critical of earlier conclusions[36,37] based on results obtained by this method. For aqueous solutions of aliphatic acids and alcohols such results confirm the importance of diffusion control, except for an initial time lag during which longer times are required for a given degree of adsorption to be reached than would be expected if diffusion were the only limiting factor.[38] The existence of a small energy barrier seems to be the most satisfactory explanation of this stage. Values of about 2 kcal./mole for iso-amyl alcohol have been calculated, falling to about 1·3 and 0·7 kcal./mole for n-hexyl and n-heptyl alcohols, respectively.[39]

Addison has measured migrational velocities for a series of aliphatic alcohols in aqueous solutions.[40] The velocity of migration V is given by

$$V = -\frac{100}{RTt}\frac{\mathrm{d}\gamma}{\mathrm{d}c},$$ (13.10)

where V is the mean velocity of migration to the surface of the solution, t is the time required for equilibrium to be established, T is the absolute temperature, and γ the surface tension at concentration c. The velocity is very dependent on the chain-length of the normal alcohols (Fig. 13.4) and some differences are found between isomers.

The true velocity is obtained from the variation of surface excess with time,[41] and typical curves are shown in Fig. 13.5. They can be fitted by the equation:

$$v = a^n t\,\mathrm{e}^{-akt},$$ (13.11)

where v is the true velocity, t is the age of the surface, and a, k, and n are the constants for a given solute. The value of n depends only on the solute. For the aliphatic alcohols it increases regularly with chain-length; k decreases with chain-length; a is dependent on concentration, the relationship depending on chain-length, but a/c has a limiting value at infinite dilution independent of chain-length.

This equation was subsequently modified to[42]:

$$\frac{\mathrm{d}\Gamma}{\mathrm{d}t} = \frac{c}{100}\,a^n t\,\mathrm{e}^{-akt}.$$ (13.12)

This governs the adsorption of long-chain as well as short-chain alcohols at the air–water interface. The times involved are longer than would correspond to diffusion control, and the existence of an energy barrier is postulated,[39] related to k in the above equation.[42] This is mainly due to the energy needed to overcome the attraction of the water molecules, the

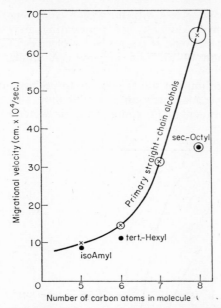

FIG. 13.4. Mean velocities of migration of alcohols to the surface of aqueous solutions as a function of chain-length.[40] (Reproduced with permission from the *Journal of the Chemical Society*.)

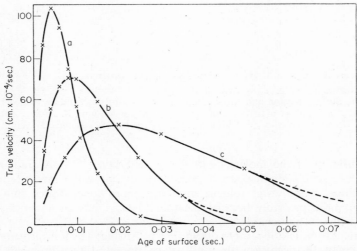

FIG. 13.5. True velocity of migration of n-heptyl alcohol to the surface of aqueous solutions[40] at concentrations of (a) 0·06%, (b) 0·04%, (c) 0·02%. (Reproduced with permission from the *Journal of the Chemical Society*.)

energy barrier becoming less with increasing chain-length. Minor contribu-
tions may come from difficulty of penetrating an existing film against
lateral interactions between the adsorbed molecules; this becomes greater
as the concentration of the film increases. A similar energy barrier to
desorption exists.[43]

In an expanding film, the solute appears to be less surface active than it
would be at the same concentration in a static film. This has been attributed,
for films of decanoic acid and decanol at the air–water interface, to a change

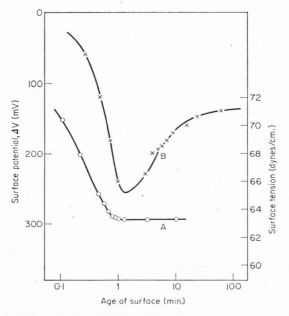

Fig. 13.6. Surface tension and surface potential of an aqueous solution of decyl alcohol
(0·00045%) at 15°C as a function of time: A, surface tension; B, surface potential.[47]
(Reproduced with permission from the *Journal of the Chemical Society*.)

in orientation of the molecules in the surface layer.[44, 45] A similar effect
is observed with sodium dodecyl sulphate at the air–water interface and for
dilute films at the water–toluene interface; more concentrated films resist
disorientation, probably because the toluene layer has an anchoring effect
on the hydrocarbon chains.[46]

Reorientation is not always detectable by surface tension measurements,
and it may be necessary to use surface potential measurements. After
aqueous solutions of decanol have reached a limiting surface tension, considerable change in surface potential still occurs (Fig. 13.6).[47] The reorienta-

tion process is not identical for different members of the homologous series of alcohols.[48]

In contrast to both of these conclusions, Hansen has found[27] that an empirical equation,

$$\frac{\pi}{\pi_\infty - \pi} e^{-b\pi/\pi_\infty} = k\pi_\infty^2 t,$$ (13.13)

where π is the surface pressure $(\gamma_0 - \gamma)$ at time t, π_∞ is the equilibrium

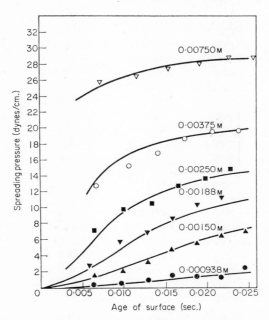

FIG. 13.7. Spreading pressure as a function of time for aqueous solutions of heptanoic acid at 20°c. Points, experimental; curves according to equation (13.9).[27] (Reproduced with the permission of the American Chemical Society from the *Journal of Physical Chemistry*.)

surface pressure for the solution, and b and k are constants, fits data for solutions of fatty acids (Fig. 13.7) and alcohols obtained by three different techniques: vibrating jet (Hansen), drop weight (Addison), and Wilhelmy plate (Dervichian). The constant b is approximately 1 in each case. The constant k varies with solubility; for the fatty acids, $\log k$ is a linear function of log (solubility).

In interpreting the above equation, Hansen drew attention specifically to the assumption that surface tension, measured at any time, corresponds to the surface excess, i.e. the surface excess of one solution, measured before equilibrium is established, is the same as that of a second solution at equilibrium which has the same surface tension. This implies that the rate-determining step in adsorption occurs before or during entry to the surface, not in reorientation or rearrangement within the surface film. With this assumption, it was shown that adsorption is too slow to be diffusion controlled, except possibly for concentrated solutions. Although no explicit alternative mechanism could be put forward, the results were consistent with a second order kinetic mechanism, implying a dimeric transition state located substantially in the surface film.

THE LIQUID–LIQUID INTERFACE

As at the liquid–vapour interface, the time required for attainment of equilibrium decreases with increase in concentration of a given solute.[49] A kinetic theory of adsorption can be based on diffusion as the rate-controlling step,[50] but diffusion alone is insufficient to explain observed results, and it appears that a barrier exists at the interface. A specific example is the adsorption of palmitic acid at the water–hexane interface. The relatively slow rate of adsorption has been attributed to the breakdown of dimeric acid (present in the hexane phase) to give monomers which could be adsorbed at the water surface. Even this, however, is inadequate quantitatively, and it is suggested alternatively that there is an entropy barrier at the surface, it being necessary for molecules to lose some of their degrees of freedom before being adsorbed.[51]

For adsorption of the straight-chain alcohols at the water–liquid paraffin interface, the magnitude of the barrier decreases with increase in chain-length of the solute molecule,[52] as for the liquid–vapour interface (see above). The rates, both of migration through the aqueous phase and of penetration into the surface layer, appear to be a function primarily of the size and structure of the hydrophobic part of the solute molecule, differences in the polar group being relatively unimportant.

For the specific case of adsorption at a mercury surface, it seems satisfactory to assume that adsorption of polymers is diffusion controlled.[53]

REFERENCES

1. G. J. Young, J. J. Chessick, and F. H. Healey, *J. Phys. Chem.*, 1956, **60**, 394.
2. E. B. Greenhill, *Trans. Faraday Soc.*, 1949, **45**, 625.
3. L. E. Swearingen and B. N. Dickenson, *J. Phys. Chem.*, 1932, **36**, 534.
4. I. M. Kolthoff and A. Kahn, *J. Phys. Chem.*, 1950, **54**, 251.
5. G. Kraus and J. Dugone, *Ind. Eng. Chem.*, 1955, **47**, 1809.
6. S. Rothman, J. Mandel, F. R. McCann, and S. G. Weissberg, *J. Colloid Sci.*, 1955, **10**, 338.
7. J. F. Hobden and H. H. G. Jellinek, *J. Polymer Sci.*, 1953, **11**, 365.
8. I. Claesson and S. Claesson, *Arkiv. Kemi, Min., Geol.*, 1945, **19**A, No. 5.
9. C. Peterson and T. K. Kwei, *J. Phys. Chem.*, 1961, **65**, 1330.
10. H. H. G. Jellinek and H. L. Northey, *J. Polymer Sci.*, 1954, **14**, 583.
11. R. N. Smith, C. F. Geiger, and C. Pierce, *J. Phys. Chem.*, 1953, **57**, 382.

12. J. H. de Boer, "The Dynamical Aspects of Adsorption", Clarendon Press, Oxford, 1953, p. 38.
13. J. W. Shepard and J. P. Ryan, *J. Phys. Chem.*, 1959, **63**, 1729.
14. H. A. Smith and T. Fort, *J. Phys. Chem.*, 1958, **62**, 519.
15. S. Lagergren, *Bil. K. Svenska Vetenskapsakad. Handl.*, 1898, **24** (2), No. 4.
16. J. Kawecka, J. Klesinska-Drwalowa, and A. K. M. Lasoń, *Chem. Stosowana*, 1963, **7**, 441.
17. A. Fava and H. Eyring, *J. Phys. Chem.*, 1956, **60**, 890.
18. J. Pouchlý and E. Erdös, *Coll. Czech. Chem. Comm.*, 1958, **23**, 1706.
19. J. Pouchlý, *Coll. Czech. Chem. Comm.*, 1959, **24**, 3007.
20. J. Pouchlý, *Coll. Czech. Chem. Comm.*, 1960, **25**, 1397.
21. R. L. Patrick and G. O. Payne, *J. Colloid Sci.*, 1961, **16**, 93.
22. W. N. Bond, *Proc. Phys. Soc.*, 1935, **47**, 549.
23. H. O. Puls, *Phil. Mag.*, 1936, [7], **22**, 970.
24. K. L. Sutherland, *Rev. Pure Appl. Chem. (Australia)*, 1951, **1**, 35.
25. R. Defay and J. R. Hommelen, *J. Colloid Sci.*, 1959, **14**, 401.
26. D. G. Dervichian, *Kolloid Z.*, 1956, **146**, 96.
27. R. S. Hansen and T. C. Wallace, *J. Phys. Chem.*, 1959, **63**, 1085.
28. C. C. Addison, *J. Chem. Soc.*, 1944, 252.
29. A. M. Posner and A. E. Alexander, *J. Colloid Sci.*, 1953, **8**, 575.
30. A. F. H. Ward and L. Tordai, *Trans. Faraday Soc.*, 1946, **42**, 413.
31. R. Defay and J. R. Hommelen, *J. Colloid Sci.*, 1959, **14**, 411.
32. H. Hotta and T. Isemura, *Bull. Chem. Soc. Japan*, 1952, **25**, 101.
33. R. Defay and T. Roba-Thilly, *J. Colloid Sci.*, 1954, **9**, Supplement I, 36.
34. F. MacRitchie and A. E. Alexander, *J. Colloid Sci.*, 1963, **18**, 453.
35. K. L. Sutherland, *Austral. J. Sci. Res.*, 1952, A, **5**, 683.
36. W. N. Bond and H. O. Puls, *Phil. Mag.*, 1937, [7], **24**, 864.
37. S. Ross and R. M. Haak, *J. Phys. Chem.*, 1958, **62**, 1260.
38. R. S. Hansen, *J. Colloid Sci.*, 1961, **16**, 549.
39. C. M. Blair, *J. Chem. Phys.*, 1948, **16**, 113.
40. C. C. Addison, *J. Chem. Soc.*, 1948, 98.
41. C. C. Addison, *J. Chem. Soc.*, 1945, 354.
42. C. C. Addison and S. K. Hutchinson, *J. Chem. Soc.*, 1949, 3387.
43. C. C. Addison and S. K. Hutchinson, *J. Chem. Soc.*, 1949, 3395.
44. C. C. Addison, J. Bagot, and H. S. McCauley, *J. Chem. Soc.*, 1948, 936.
45. C. C. Addison and S. K. Hutchinson, *J. Chem. Soc.*, 1949, 3406.
46. C. C. Addison and S. K. Hutchinson, *J. Chem. Soc.*, 1948, 943.
47. C. C. Addison and D. Litherland, *J. Chem. Soc.*, 1953, 1143.
48. C. C. Addison and D. Litherland, *J. Chem. Soc.*, 1953, 1159.
49. F. H. Garner and P. Mina, *Trans. Faraday Soc.*, 1959, **55**, 1607.
50. A. F. H. Ward and L. Tordai, *J. Chem. Phys.*, 1946, **14**, 453.
51. A. F. H. Ward, "Surface Chemistry", Butterworths, London, 1949, p. 55.
52. F. H. Garner and P. Mina, *Trans. Faraday Soc.*, 1959, **55**, 1616.
53. I. R. Miller, *Trans. Faraday Soc.*, 1961, **57**, 301.

CHAPTER 14

Thermodynamics of Adsorption from the Liquid Phase

INTRODUCTION

The thermodynamic treatment of adsorption from solution involves a satisfactory treatment of the solution itself as well as that of the interfacial phenomena. It is therefore more complex than that of adsorption of single vapours by solids which has received consistent attention only in comparatively recent years. The difference between adsorption of a single vapour and adsorption from a solution has been emphasized in previous chapters. This remains relevant in considering such terms as "heat of adsorption". For adsorption from solution, this is essentially a heat of transfer of one component from the bulk phase to the interface. It thus depends on the initial environment of the component being considered. For dilute solutions, this can be considered to be approximately constant for all surface coverages. When the complete concentration range for binary liquid mixtures is considered, however, the initial environment varies very considerably. Some care is therefore desirable in defining the precise significance of such terms in relation to adsorption from solution.

At present it seems very desirable to recognize, as has been emphasized in previous chapters, the similarity between the solution–solid interface, the solution–vapour interface, and the different types of liquid–liquid interface. Very few papers have been written with this in mind, and there is a considerable lack of coherence in any summary of the existing literature. The surface tension of liquids has recently been considered from the standpoint of thermodynamics and statistical thermodynamics by Ono and Kondo.[1]

ADSORPTION AT THE LIQUID–VAPOUR INTERFACE

Butler pointed out[2] that, in Gibbs' thermodynamic treatment of interfaces, the absence of a model—specifically a model for molecular arrangement—limited the determination of the composition of surface layers. Such further progress as can be made is limited to the validity of the model which is used. Butler examined a model in which the adsorbed phase was limited to a single layer of molecules. This at present seems to be a useful model for many, though not for all systems. A claim[3] that it is incompatible with the Gibbs adsorption equation has been disputed.[4]

With this model, it can be shown that the chemical potentials (or partial molar free energies) of components in the surface layer (μ_1^s, μ_2^s, ..., μ_i^s) are related to those of the components in the bulk liquid (μ_1, μ_2, ..., μ_i) by the equation:

$$\frac{\mu_1^s - \mu_1}{A_1} = \frac{\mu_2^s - \mu_2}{A_2} = ... \frac{\mu_i^s - \mu_i}{A_i} = \gamma, \tag{14.1}$$

where A_1, A_2, ..., A_i are the corresponding partial molar surface areas of the respective components, and γ is the surface tension of the solution. In general,

$$\mu_1 = \mu_1^0 + RT \ln a_1 \tag{14.2}$$

and

$$\mu_1^s = (\mu_1^s)^0 + RT \ln a_1^s, \tag{14.3}$$

where μ_1^0 and $(\mu_1^s)^0$ refer to the chemical potentials in standard states, and a_1 and a_1^s to the activities in the bulk phase and the surface layer respectively. As in treatments of homogeneous systems the choice of standard state has depended upon whether the author has been concerned with dilute solutions or the whole range of composition for completely miscible liquids.

From the above equations it follows that:

$$\ln a_1^s - \ln a_1 = \frac{1}{RT} [\gamma A_1 - \{(\mu_1^s)^0 - \mu_1^0\}] \tag{14.4}$$

or

$$a_1^s = a_1 \, e^{[\gamma A_1 - \{(\mu_1{}^s)^0 - \mu_1{}^0\}]/RT}. \tag{14.5}$$

Further assumptions are now needed before equation (14.4) can be used.

Dilute Solutions

For dilute solutions, Butler has shown that Szyszkowski's equation can be derived if a number of assumptions are made. Activities are defined so that $a_1 = 1$ when $x_1 = 1$, and $a_2 = 1$ when x_2 is small (x_1 and x_2 being the corresponding mole fractions), with corresponding definitions for the surface layer. For the pure solvent, $a_1 = 1$ and $a_1^s = 1$. Hence from equation (14.5) it follows that

$$[(\mu_1^s)^0 - \mu_1^0]/A_1 = \gamma_0, \tag{14.6}$$

where γ_0 is the surface tension of the pure solvent. Equation (14.4) can then be written as

$$\ln a_1^s - \ln a_1 = \frac{A_1}{RT} (\gamma - \gamma_0). \tag{14.7}$$

If the two components have the same partial molar surface area (i.e. $A_1 = A_2 = A$), then from equation (14.4) and the corresponding equation for the second component:

$$\ln \frac{a_2^s}{a_2} - \ln \frac{a_1^s}{a_1} = \frac{1}{RT} [\{(\mu_1^s)^0 - \mu_1^0\} - \{(\mu_2^s)^0 - \mu_2^0\}]. \tag{14.8}$$

If $a_1 = x_1 f_1$, $a_2 = x_2 f_2$, etc., where the terms in f are activity coefficients, then from (14.8) it follows that

$$\frac{x_2^s}{x_1^s} = \frac{x_2 f_2 f_1^s}{x_1 f_1 f_2^s} \, e^{\frac{1}{RT}[\{(\mu_1^s)^0 - \mu_1^0\} - \{(\mu_2^s)^0 - \mu_2^0\}]}.$$

If f_1^s/f_2^s is constant, this can be written as:

$$\frac{x_2^s}{x_1^s} = \frac{x_2 f_2}{x_1 f_1} \frac{1}{K},$$

where K is a constant, and equals $e^{(\Delta\mu_2 - \Delta\mu_1)/RT}$. As $x_1^s + x_2^s = 1$,

$$x_1^s = \frac{1}{1 + \dfrac{x_2 f_2}{x_1 f_1} \dfrac{1}{K}} = \frac{K x_1 f_1}{K x_1 f_1 + x_2 f_2}. \tag{14.9}$$

From equations (14.9) and (14.7) we have:

$$\ln \left[\frac{f_1^s}{a_1 \left(1 + \dfrac{x_2 f_2}{x_1 f_1}\right) \dfrac{1}{K}} \right] = \frac{A}{RT}(\gamma - \gamma_0).$$

$$\therefore \quad \ln \left[\frac{f_1^s}{a_1 + \dfrac{x_2 f_2}{K}} \right] = \frac{A}{RT}(\gamma - \gamma_0).$$

For sufficiently dilute solutions (x_2 small) $a_1 \sim 1$. If it is also assumed that $f_1^s \sim 1$, then

$$\frac{A}{RT}(\gamma - \gamma_0) = \ln \left(\frac{1}{1 + \dfrac{x_2}{K}} \right)$$

or

$$\frac{\gamma}{\gamma_0} = 1 - \frac{RT}{A\gamma_0} \ln \left(\frac{x_2 + K}{K} \right), \tag{14.10}$$

which has the same form as equation (11.23).

Butler examined the application of equation (14.10) to dilute aqueous solutions of n-butyl alcohol, n-butyric acid, and phenol. Reasonable agreement was found for values of a_2 between 0·1 and 0·5. In each case the value of A was 26·6 sq. Å. This is of the right order of magnitude, and is perhaps as good a value as could be expected in view of the assumption made in deriving equation (14.10). For the systems in question it is clearly not the case that $A_1 = A_2$.

Concentrated Solutions

A development of the above treatment for the simpler binary mixtures of completely miscible liquids has been put forward by several authors.[5-8] Equation (14.7) can be used in the form:

$$\gamma = \gamma_1 + \frac{RT}{A_1} \ln \frac{a_1^s}{a_1} = \gamma_2 + \frac{RT}{A_2} \ln \frac{a_2^s}{a_2}, \tag{14.11}$$

where γ_1 and γ_2 are the surface tensions of components 1 and 2, respectively. For the simplest case it is assumed that substances which mix ideally in the liquid phase also mix ideally in the surface layer. If, also, $A_1 = A_2 = A$, then from equation (14.11) we have

$$\frac{x_2^s x_1}{x_1^s x_2} = e^{(\gamma_1 - \gamma_2)A/RT}, \tag{14.12}$$

whence

$$x_1^s = \frac{x_1}{x_1 + cx_2}, \tag{14.13}$$

where

$$c = e^{(\gamma_1 - \gamma_2)A/RT}. \tag{14.14}$$

Thus

$$\gamma = \gamma_1 - \frac{RT}{A} \ln [1 - (1 - c)x_2]. \tag{14.15}$$

Schuchowitzky has found this equation to apply well to mixtures of ethyl ether and benzene at 18°c, and to benzene and nitrobenzene at 25°c, A in each case being taken as 28 sq. Å.

If the two components have different molecular areas, an exact equation for γ in terms of x_2 is not obtained, but a first approximation has been found successful[5] in describing the benzene–carbon disulphide system at 18°c. In this case the experimental data are fitted by using a molecular area of 8 sq. Å for carbon disulphide. This implies that the molecule is adsorbed with the major axis perpendicular to the surface.

If c is replaced by $\frac{1}{K}$, where $K = e^{(\gamma_2 - \gamma_1)A/RT}$, then equation (14.13) becomes

$$x_1^s = \frac{Kx_1}{Kx_1 + x_2}, \tag{14.16}$$

and is then of the same form as (14.9), except for omission of the activity coefficients.

A treatment which allows for non-ideality in the adsorbed layer has been put forward by Stauff.[9] This is found to be quite effective for some, but not all, aqueous systems. An analogous treatment, which also allows for non-ideality in the liquid phase, is described below (p. 252).

ADSORPTION AT THE LIQUID–SOLID INTERFACE

A simple treatment, analogous to that given in the preceding sections, helps to show how the nature of the solution determines the form of the isotherm in adsorption onto a solid surface from completely miscible liquids. It has been developed for ideal and for regular solutions;[10] cf. Erdös.[11] For each case it is assumed that the change in chemical potential for the pure component arising from adsorption is $\Delta\mu^0$. Thus:

$$(\mu_1^s)^0 = \mu_1^0 + \Delta\mu_1^0, \tag{14.17}$$

and for binary mixture

$$\mu_1^0 + RT \ln a_1 = \mu_1^0 + \Delta\mu_1^0 + RT \ln a_1^s, \tag{14.18}$$

$$\mu_2^0 + RT \ln a_2 = \mu_2^0 + \Delta\mu_2^0 + RT \ln a_2^s. \tag{14.19}$$

Ideal Solutions

For an ideal mixture, activities can be replaced by mole fractions, and it follows from equations (14.18) and (14.19) that:

$$\ln \frac{x_1^s}{1-x_1^s} = \frac{\Delta\mu_2^0 - \Delta\mu_1^0}{RT} + \ln \frac{x_1}{1-x_1}, \tag{14.20}$$

or

$$\frac{x_1^s}{1-x_1^s} = \frac{x_1}{1-x_1} e^{\frac{\Delta\mu_2{}^0 - \Delta\mu_1{}^0}{RT}}. \tag{14.21}$$

(If the treatment given by Butler for the liquid–vapour interface (see above) were followed, equation (14.20) would contain a further term $+\gamma(A_1 - A_2)/RT$ on the right-hand side; γ is the interfacial tension and A_1 and A_2 the respective partial molar surface areas. This term becomes zero if the two components have molecules which occupy the same area at the surface.) Hence the individual isotherm for component 1, expressed in terms of its mole fraction in the adsorbed layer ($x_1^s = n_1^s/n^s$), is given by

$$x_1^s = \frac{x_1\, e^{(\Delta\mu_2{}^0 - \Delta\mu_1{}^0)/RT}}{1 + x_1[e^{(\Delta\mu_2{}^0 - \Delta\mu_1{}^0)/RT} - 1]}. \tag{14.22}$$

If $e^{(\Delta\mu_2{}^0 - \Delta\mu_1{}^0)/RT} = K$, then

$$x_1^s = \frac{Kx_1}{1 + (K-1)x_1} = \frac{Kx_1}{Kx_1 + x_2}, \tag{14.23}$$

which has the same form as equation (14.16) for the liquid–vapour interface. As, from equation (3.4),

$$\frac{n_0 \Delta x}{m} = n^s(x_1^s - x_1),$$

it follows[11] that:

$$\frac{n_0 \Delta x}{m} = n^s \left[\frac{Kx_1}{Kx_1 + x_2} - x_1 \right]. \tag{14.24}$$

This can be put into a linear form:

$$\frac{n_0 \Delta x}{x_2 m} = n^s \left(\frac{K-1}{K}\right) - \frac{n_0 \Delta x}{m} \frac{1}{Kx_1}, \tag{14.25}$$

which is a useful form for testing the equation. Šišková and Erdös have found that equation (14.25)† is obeyed over a wide range of concentration for adsorption from a number of mixtures by silica gel (e.g. carbon tetrachloride–benzene) though it is not satisfactory for all mixtures and tends to be less satisfactory for adsorption by activated charcoal. There are two consequences for the ideal system. The first, which follows from equation (14.21), is that preferential adsorption of one component must take place at all concentrations. The second is that the individual isotherm (in terms of mole fractions) follows an equation of the Langmuir type; equation (14.22) has the same form as equation (4.9).

From a more detailed analysis, Everett[12] has derived the equation:

$$\frac{x_1^s}{x_2^s} = K_1 \frac{x_1}{x_2}, \tag{14.26}$$

from which it follows that:

$$x_1^s = \frac{K_1 x_1}{K_1 x_1 + x_2}, \tag{14.27}$$

where

$$K_1 = e^{\{[\Delta_a U_2 - \Delta_a U_1]/RT - [\Delta_a S_2^* - \Delta_a S_1^*]/R\}},$$

in which $\Delta_a U$ is the change in energy and $\Delta_a S^*$ is the change in the thermal part of the entropy on adsorption.

As

$$\frac{n_0 \Delta x}{m} = n^s(x_1^s - x_1)$$

for any system, it follows from equation (14.23) that:

$$\frac{n_0 \Delta x}{m} = \frac{n^s (K_1 - 1) x_1 x_2}{K_1 x_1 + x_2}. \tag{14.28}$$

Further

$$n_1^s = \frac{n^s K_1 x_1}{K_1 x_1 + x_2}, \tag{14.29}$$

which is equivalent to equation (4.11).

Rearrangement of (14.28) gives:

$$\frac{x_1 x_2}{n_0 \Delta x / m} = \frac{1}{n^s} \left[x_1 + \frac{1}{1 - K_1}\right]. \tag{14.30}$$

Thus[12] n^s and K_1 can be calculated from the slope and intercept of a linear

† In the original form,[11] this equation was expressed in terms of a constant which is the reciprocal of K.

plot—for systems to which the assumptions underlying equation (14.30) apply. Equations (14.25) and (14.30) are alternative linear forms of the same equation for $n_0 \Delta x/m$. The former is probably the better for evaluating K, and the latter for evaluating n^s.

For a series of such systems with different values of K_1, a family of isotherms of concentration change can be drawn (Fig. 14.1). The maximum value of $n_0 \Delta x/m$ in each case is[11]

$$n^s \left(\frac{\sqrt{K_1} - 1}{\sqrt{K_1} + 1} \right),$$

for a value of

$$x_1 = \frac{1}{\sqrt{K_1} + 1}.$$

[Šišková and Erdös[11] give equivalent expressions in terms of a constant which is $1/K_1$.] Thus

$$\left(\frac{n_0 \Delta x}{m} \right)_{max} = n^s (1 - 2x_{max}), \tag{14.31}$$

i.e. there is a straight line through the maxima of a family of isotherms. This is shown in the curves calculated by Everett[12] (Fig. 14.1).

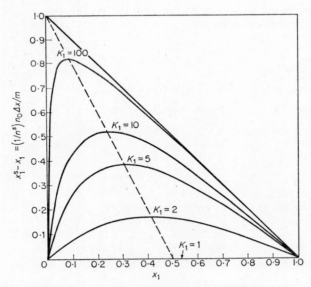

FIG. 14.1. Isotherms of surface excess[12] calculated according to equation (14.26) for various values of K_1. (Reproduced with permission from *Transactions of the Faraday Society*.)

Equation (14.14) can be used to relate adsorption at the liquid–vapour interface to adsorption at the liquid–solid interface for an ideal system, by means of Young's equation:

$$\gamma_{LS} = \gamma_{SV} - \gamma_{LV} \cos\theta,$$

where the subscripts L, S, and V refer to liquid, solid, and vapour, respectively. Thus:

$$(\gamma_{LS})_1 - (\gamma_{LS})_2 = \gamma_{SV} - (\gamma_{LV})_1 \cos\theta_1 - \gamma_{SV} + (\gamma_{LV})_2 \cos\theta_2.$$

If

$$\theta_1 = \theta_2 = 0^\circ,$$

then

$$(\gamma_{LS})_1 - (\gamma_{LS})_2 = (\gamma_{LV})_2 - (\gamma_{LV})_1,$$

and

$$c_{LS} = \frac{1}{c_{LV}}, \qquad (14.32)$$

a relation which must apply to any liquid–solid interface at which there is ideal behaviour.

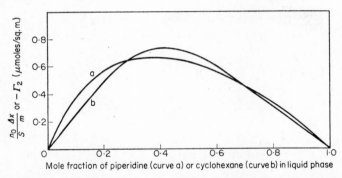

Fig. 14.2. Isotherms of surface excess for the system cyclohexane-piperidine: (a) at the liquid–Graphon interface, (b) at the liquid–vapour interface.[13]

It follows from equation (14.13) that if the "surface excess" isotherms are plotted in terms of unit area of surface, the curve for any liquid–solid interface (for which these conditions apply) can be obtained by rotating that for the liquid–vapour interface through 180°. An example is shown in Fig. 14.2 of the closeness with which this can be observed in practice, even though the liquid mixture (cyclohexane–piperidine) is not ideal.[13]

More precisely,[12] in Everett's nomenclature,

$$\ln\frac{K_1}{K_4} = -(\Delta_a U_1 - \Delta_a U_2)/RT, \qquad (14.33)$$

where

$$K_4 = e^{(\gamma_2{}^0 - \gamma_1{}^0)A/RT}$$

for the liquid–vapour interface. Equation (14.33) is based on the assumption that the thermal entropies at the two interfaces are equal and that, approximately:

$$(\Delta_a U_1)_{LS} = (\Delta_a U_1)_{LV} + (\Delta_a U_1)_{VS}. \tag{14.34}$$

Regular Solutions

For regular solutions, Schay[10] uses activity coefficients defined by

$$\ln f_1 = q(1-x_1)^2 \quad \text{and} \quad \ln f_2 = qx_1^2,$$

where q is a measure of the interaction between the two components at a given temperature. A different term, q', is required for the adsorbed layer. One way of relating q' to q is to assume that the adsorbed layer is a monolayer, and that the molecules can be regarded as cubes, of the same size for the two components. Then each molecule in the bulk phase has 26 nearest neighbours (including those with common corners), whereas each molecule in the surface layer has only 17. Hence $q' = \dfrac{17q}{26} = 0.654q$. Then:

$$RT \ln x_1 + RTq(1-x_1)^2 = \Delta\mu_1^0 + RT \ln x_1^s + RTq'(1-x_1^s)^2, \tag{14.35}$$

$$RT \ln(1-x_1) + RTqx_1^2 = \Delta\mu_2^0 + RT \ln(1-x_1^s) + RTq'(x_1^s)^2. \tag{14.36}$$

$$\therefore \quad RT \ln \frac{x_1}{1-x_1} + RTq(1-2x_1) = \Delta\mu_1^0 - \Delta\mu_2^0 +$$

$$RT \ln \frac{x_1^s}{1-x_1^s} + RTq'(1-2x_1^s), \tag{14.37}$$

or

$$\ln \frac{x_1^s}{1-x_1^s} - 1.308qx_1^s = \ln \frac{x_1}{1-x_1} + \frac{\Delta\mu_2^0 - \Delta\mu_1^0}{RT} + 0.346q - 2qx_1. \tag{14.38}$$

On this basis, a reversal of the sign of preferential adsorption [which is related to $(x_1^s - x_1)$, cf. equation (3.4)] is possible between $x = 0$ and $x = 1$, i.e. an S-shaped composite isotherm is possible. This can only occur, however, if the difference between $\Delta\mu_2^0$ and $\Delta\mu_1^0$ is small; otherwise, the more strongly adsorbed component is preferentially adsorbed over the whole of the concentration range.

The criterion given by Erdős,[14] if applied to Schay's results, leads to the conclusion that an S-shaped isotherm is obtained only if $(\Delta\mu_2^0 - \Delta\mu_1^0)/RT <$ $0.345/q$. The term $(\Delta\mu_2^0 - \Delta\mu_1^0)/RT$ represents the extent to which component 1 is more stable than component 2 in the adsorbed phase, relative to the bulk solution.

In terms of a physical model for the regular solution, each component has a tendency to "escape" from a dilute solution; this provides the possibility for each to be adsorbed preferentially from a dilute solution. If, however, the interaction of the surface with one component is much greater than with the other, this may overcome the effect due to the "escaping tendency" and result in preferential adsorption of the first component at all concentrations.

Schay has illustrated the importance of these two effects for mixtures of benzene and ethanol, using a value of q derived from vapour pressure data. If the strengths of interaction of the two components with the surface are equal (i.e. $\Delta\mu_2^0 = \mu\Delta_1^0$), curve I in Fig. 14.3 is obtained. If there is a difference

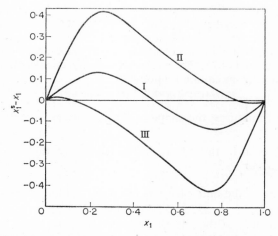

FIG. 14.3. Calculated isotherms for adsorption from mixtures of benzene and ethanol.[10] (Reproduced with permission from *Acta Chimica Acad. Sci. Hung.*)

of $0.5RT$ in the values of $\Delta\mu^0$, then curve II or curve III is obtained, depending on which component is the more strongly adsorbed.

This treatment is probably not perfectly precise as the molecules of benzene and ethanol are unlikely to occupy equal areas at the surface. It would be difficult to take into account the effect of the extra term, $\gamma(A_1 - A_2)/RT$, as data for the interfacial (liquid–solid) tensions are not readily available. A simple calculation, however, shows that it is probably significant for real systems. Thus if the interfacial tension is taken to be of the same order of magnitude as the adhesion tension, for which a representative value[15] is 60 ergs/sq. cm., and the difference in surface molecular area of the two components is taken to be 15 sq. Å, then $\gamma(A_1 - A_2)$ would equal about $2RT$.

Statistical mechanical treatments of this problem have been put forward by Elton[16] and by Delmas and Patterson.[17] Elton has also suggested[18] that the term $[\mu_1^0 - (\mu_1^s)^0]$ which equals $-\Delta\mu_1^0$ in Schay's nomenclature, should be referred to as the adsorption potential of component 1. His prediction that complete preferential adsorption will rarely be found is not entirely in accordance with the above analysis, though once more the effect of interfacial tension should be taken into account. In practice, many cases of complete preferential adsorption are now known, and this seems to be the normal situation unless the surface is heterogeneous or, in some cases, the binary mixture is far from ideal (cf. Chapter 10).

In general, the precise form of the equation for regular solutions, analogous to equation (14.13) or (14.22) for ideal solutions, is:

$$x_1^s = \frac{x_1}{x_1 + x_2 \dfrac{f_2 f_1^s}{f_1 f_2^s} e^{(\gamma_1 A_1 - \gamma_2 A_2)/RT} \cdot e^{\gamma(A_2 - A_1)/RT}}. \tag{14.39}$$

For mixtures which are ideal in both surface phase and bulk liquid, this reduces to equation (14.22) because each activity coefficient is unity, and $A_1 = A_2$; further, from equation (14.6),

$$e^{(\gamma_1 A_1 - \gamma_2 A_2)/RT} = e^{(\Delta\mu_1^0 - \Delta\mu_2^0)/RT}.$$

For non-ideal mixtures, the shape of the composite isotherm depends on four factors[19-20]:

(i) the difference in the adsorption potentials of the two components, $e^{(\gamma_1 A_1 - \gamma_2 A_2)/RT}$,

(ii) the nature of the bulk phase as given by $\dfrac{f_2}{f_1}$,

(iii) the interaction of the adsorbed species, as given by $\dfrac{f_1^s}{f_2^s}$,

(iv) the variation of the surface free energy with composition of the bulk phase, as given by $e^{\gamma(A_2 - A_1)/RT}$.

The importance of the second factor is often quite considerable.[20]

On the particular question as to whether a given system gives a U-shaped or S-shaped isotherm it has been shown that an S-shaped isotherm can be produced for adsorption at the liquid–vapour interface if the curve of γ against x passes through a maximum or minimum.[19] For the liquid–solid interface, values of γ were obtained by graphical integration on the basis of the Gibbs equation (cf. equations 11.8 and 11.15):

$$\frac{n_0 \Delta x}{Sm} = \Gamma_1^{(N)} = -\frac{x_2}{RT} \frac{d\gamma}{d\ln a_1}.$$

Hence

$$(\gamma_1 - \gamma)A_1 = \frac{RTA_1}{S} \int_{a_1 = 1}^{a_1} \frac{n_0 \Delta x}{x_2 m} \cdot d\ln a_1. \tag{14.40}$$

This gives a further basis for comparing the liquid–solid and liquid–vapour interfaces, as the variation of γ with x can be measured directly for the latter.

Surface Activity Coefficients

The importance of the activities of adsorbed species is becoming clearer and a few attempts have been made to evaluate surface activity coefficients. Everett[21] has pointed out that some of these[18,22,23] have given incorrect values, essentially because the γA term has been omitted from equations analogous to (14.4).

Everett[21] has derived a rigorous method for calculating surface activity coefficients for simple systems in which the adsorbed phase is confined to a single molecular layer and the two components (A and B) occupy the same molecular area. An exchange reaction:

$$A_{liquid} + B_{surface} \rightleftarrows A_{surface} + B_{liquid} \tag{14.41}$$

can then occur without change in the interfacial area, and the condition for equilibrium is:

$$\mu_1 + \mu_2^s = \mu_2 + \mu_1^s, \tag{14.42}$$

whence

$$\frac{x_1^s f_1^s x_2 f_2}{x_1 f_1 x_2^s f_2^s} \, e^{\{[\mu_1{}^0 - (\mu_1{}^s)^0] - [\mu_2{}^0 - (\mu_2{}^s)^0]\}/RT} = K, \tag{14.43}$$

where K is a constant for a given temperature. It can be shown that

$$\ln K = \int_0^1 \ln \left(\frac{x_1^s x_1 f_1}{x_2^s x_2 f_2} \right), \tag{14.44}$$

and that this value of K can be used to evaluate f_1^s and f_2^s. This has been done for one system.

For the more complex systems, Schay uses graphical integration to obtain $(\gamma_1 - \gamma)$ and $(\gamma_2 - \gamma)$ from equation (14.40) and hence can calculate values of f_1^s and f_2^s from equation (14.39). Examples of the variation of surface activity coefficient with composition of the liquid mixture are shown[24] in Fig. 14.4.

If the two components have different molecular sizes, an extra term, r, expressing the size ratio is introduced into equation (14.43), which becomes

$$\left(\frac{a_1^s}{a_1} \right) \left(\frac{a_2}{a_2^s} \right)^r = K. \tag{14.45}$$

As r appears as a power to which two of the terms in the left-hand side of the equation are raised, the equation then becomes much more difficult to use. If there is a considerable difference in the molecular sizes, it is more difficult to define an exact boundary between the adsorbed phase and the bulk liquid.

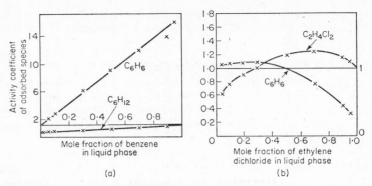

FIG. 14.4. Surface activity coefficients in adsorption from mixtures of: (a) benzene and cyclohexane on carbon black, (b) benzene and ethylene dichloride on activated charcoal.[24] (Reproduced with permission from the *Journal de Chimie Physique*.)

Equation (14.45) can be used if simplifications are introduced.[25] For ideal solutions, activities are replaced by mole fractions, whence

$$\frac{x_1^s}{x_1}\left(\frac{1-x_1}{1-x_1^s}\right)^r = K.$$

For dilute solutions, $1 - x_1 \sim 1$, so that

$$\frac{x_1^s}{x_1} \sim K(1-x_1^s)^r.$$

From a series expansion,

$$(1-x_1^s)^r \sim 1 - rx_1^s,$$

whence

$$\frac{x_1^s}{x_1} = K(1-rx_1^s),$$

and

$$x_1^s = \frac{Kx_1}{1+rKx_1}, \tag{14.46}$$

which is an equation of the Langmuir form. Its use is restricted to dilute solutions, if activity coefficients are to be taken as equal to unity, and probably to very dilute solutions if x_1^s is to be small.

Using a quasi-crystalline model for perfect solutions, Elton[16] calculated that the surface activity coefficient of a substance adsorbed at the solid–liquid interface should be a linear function of its mole fraction in the liquid phase:

$$f_1^s = (K+x_2)/K, \tag{14.47}$$

where K is an empirical constant. Equation (14.47) has not yet been tested, as the necessary experimental measurements have not been made. They would, in all probability, be difficult to obtain; the preferential adsorption from perfect solutions would probably be small and special methods of analysis might be needed for determining changes in concentration, as the two components would probably have closely similar physical properties. The system ethylene dichloride–benzene, although it comes close to obeying Raoult's Law, is not strictly an ideal solution; it does not obey equation (14.43) when adsorption takes place on charcoal.

COMPARISON OF THERMODYNAMIC AND KINETIC THEORIES OF ADSORPTION

Very recently a kinetic theory of adsorption has been formulated.[26] This is based on a lattice model for the liquid state with a corresponding lattice structure for the adsorbent. By using appropriate interatomic potential functions, the authors derived expressions for the rates of diffusion of molecules between adjacent layers and hence arrived at equations for the adsorption isotherm in which the important parameters related to the energies of interaction between molecules in the liquid state and between the adsorbent and the adsorbates.

Composite isotherms were obtained very similar in shape to those of Fig. 14.1 for the same type of system. A detailed comparison of the thermodynamic and kinetic approaches to the problem has not yet been made, but is likely to be an important future development. It will also be important to explore how far the kinetic theory can be developed to deal with systems other than those of the simplest kind.

HEAT OF ADSORPTION

The heat change which occurs when a solution is brought into contact with a solid is not as simple to consider as the heat of adsorption evolved when a single gas is adsorbed by a solid. Various meanings have been given to the term "heat of adsorption from solution", which has sometimes been used ambiguously.

A useful analysis has been made by Young, Chessick, and Healey.[27] They refer to the observed heat change as the heat of immersion, ΔH_I. This can be regarded as due to the operation of two processes. The solid comes into contact with the adsorbed layer, and a heat of wetting, ΔH_W, is involved. The adsorbed layer, however, is usually of different composition from the bulk solution, and a further heat change occurs as a result of changing the composition of the bulk liquid. This can be referred to as heat of dilution, ΔH_D, with respect to one component. Then:

$$\Delta H_I = \Delta H_W + \Delta H_D. \tag{14.48}$$

Of these terms, ΔH_D can be obtained from heats of mixing for a binary system. The heat of wetting can be calculated if an appropriate model of adsorption is used, as has also been discussed by Kiselev.[28]

Young, Chessick, and Healey assumed that molecular interactions in the adsorbed phase are approximately the same as those in the corresponding solution. The heat of wetting is then, for a unimolecular layer, the sum of the enthalpy change for covering the solid surface with the adsorbed layer, and the enthalpy change for bringing this adsorbed layer into contact with the bulk solution. If a fraction, θ, of the surface is covered with component 1, and the remainder, $(1-\theta)$, is covered with component 2, the first enthalpy change is, for a uniform surface,

$$\theta(h_s - h_{s1}) + (1-\theta)(h_s - h_{s2}),$$

where h_s is the surface enthalpy of the solid, and h_{s1} and h_{s2} the surface enthalpies of the solid with adsorbed component 1 and adsorbed component 2, respectively. The second enthalpy change is

$$(h_{s1} - h_{1\sigma}) + (1-\theta)(h_{s2} - h_{2\sigma}),$$

where $h_{1\sigma}$ and $h_{2\sigma}$ are the enthalpies of adsorbed layers of component 1 and component 2, respectively, in contact with the solution. Thus, in abbreviated nomenclature:

$$\Delta H_w = \theta\Delta h_{s1} + (1-\theta)\Delta h_{s2} + \theta\Delta h_{1\sigma} + (1-\theta)\Delta h_{2\sigma}. \qquad (14.49)$$

The terms Δh_{s1} and Δh_{s2} can be equated with the net integral energies of adsorption of the respective components, i.e. the respective differences between the energy in the adsorbed state and in the pure liquid.

The value of ΔH_w can be obtained as follows. Δh_{s1} is the difference in heats of immersion in liquid 1 of the dry solid and the solid covered with a complete layer of component 1. $\Delta h_{1\sigma}$ is the heat of immersion of solid covered with a layer of component 1 in a solution in equilibrium with this adsorbed layer; it is then assumed that this term does not vary appreciably with the concentration of the solution.

The effectiveness of this treatment for one particular system is shown in Fig. 14.5, for butanol–water–Graphon.[27] The points were obtained from

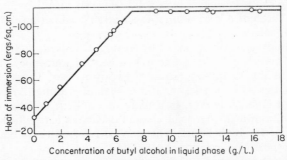

FIG. 14.5. Heats of adsorption onto Graphon from solutions of butanol in water.[27] (Reproduced with the permission of the American Chemical Society from the *Journal of Physical Chemistry*.)

experimental values of ΔH_I and ΔH_D; the line was calculated from equation (14.49). The heat of wetting becomes constant at the concentration at which it appears (from the composite isotherm) that a complete monolayer of butanol is formed. The authors derived the above treatment specifically for this system. It could be used for many more for which the above assumptions are reasonable, and could be modified for others, e.g. those involving a solid with a heterogeneous surface.

Two calorimetric methods have been used for a different measurement.[29] In one, the heats of wetting (which would be heats of immersion as defined above) of the solid by the solution and by the pure solvent were measured, and the difference was taken as the heat of adsorption of the solute. The experiment was done with a large volume of solution, to avoid a heat change due to dilution of the solution on adsorption of the solute. In the second experiment, the adsorbent was wetted with solvent before being placed in the calorimeter, and the heat of wetting was measured subsequently. The measurement appears to give

$$(1-\theta)[\Delta h_{s1} - \Delta h_{s2} + \Delta h_{1\sigma} - \Delta h_{2\sigma}] - \Delta H_D,$$

the experimental conditions being such that the last term is negligible. Thus both methods give essentially a heat of displacement (of solvent by solute).

A further method is based on injecting a known quantity of solute into a stream of solvent flowing through a column of adsorbent held in a calorimeter; the heat evolved on adsorption is then measured,[30] and is again a heat of displacement.

An isosteric heat of adsorption, comparable with ΔH_I has been obtained by using the expression:

$$\left(\frac{\partial \ln a}{\partial (1/T)}\right)_\Gamma \sim \frac{Q}{R}, \tag{14.50}$$

where a is the activity of the solution in equilibrium with an adsorbed layer of given surface excess, and Q is the heat of adsorption. Values of Q of 2·0 and 1·7 kcal./mole have been found for adsorption of naphthalene and α-methyl naphthalene, respectively, onto silica gel from n-heptane.[31]

Free Energy of Adsorption

Several authors have used the equation

$$\Delta G = -RT \ln c_s/c_1 \tag{14.51}$$

to obtain the free energy of adsorption of one component from a solution; c_s is its concentration in the surface layer, and c_1 its concentration in the bulk phase. A value of about -3 kcal./mole was obtained in this way for the adsorption of octadecyl alcohol from solutions in benzene on a number of metal powders.[32] A slightly higher value was found for its adsorption on alumina, and a value of over -6 kcal./mole for the adsorption of phenol, also from benzene, on alumina.[29] The values of c_s used in equation (14.51) depend on the thickness assumed for the adsorbed layer, and hence on the assumed orientation and configuration of the adsorbed molecules. Moreover, it is assumed in equation (14.51) that concentrations may replace activities. This is probably a good approximation for many dilute solutions, but may well be far from satisfactory for the adsorbed phase, as experimental results tend to refer to high fractional coverages of the surface.

Differential Heat of Adsorption

The differential heat of adsorption can be derived from the Gibbs free energy of adsorption by use of the equation:

$$\Delta H = \Delta G - T\left(\frac{\partial \Delta G}{\partial T}\right)_p.$$ (14.52)

Crisp[29] has used a form of the Langmuir equation,

$$\theta = Kc(1-\theta),$$

to relate c_s and c_1. This gives:

$$\Delta G = -RT \ln \frac{K}{A_0 \tau},$$ (14.53)

where τ is the thickness of the adsorbed layer, and A_0 is the limiting area of the adsorbed molecules.

This treatment is limited to adsorption from dilute solutions. Values for the heat of adsorption of aliphatic alcohols from solution in benzene by alumina were obtained in this way and also calorimetrically, as described above.[29] The results were not in striking agreement. They fell in the range 5–10 kcal./mole.

REFERENCES

1. S. Ono and S. Kondo, "Handbuch der Physik", Springer-Verlag, Berlin, Göttingen, and Heidelberg, 1960, Vol. 10, p. 134.
2. J. A. V. Butler, *Proc. Roy. Soc.*, 1932, A **135**, 348.
3. R. Defay and I. Prigogine, *Trans. Faraday Soc.*, 1950, **46**, 199.
4. T. P. Hoar and D. A. Melford, *Trans. Faraday Soc.*, 1957, **53**, 315.
5. A. Schuchowitzky, *Acta Physicochim. U.R.S.S.*, 1944, **19**, 176.
6. J. W. Belton and M. G. Evans, *Trans. Faraday Soc.*, 1945, **41**, 1.
7. E. A. Guggenheim, *Trans. Faraday Soc.*, 1945, **41**, 150.
8. J. H. Hildebrand and R. L. Scott, "The Solubility of Nonelectrolytes", Reinhold, New York, 1950, 3rd edition.
9. J. Stauff, *Z. phys. Chem. (Frankfurt)*, 1957, **10**, 24.
10. G. Schay, *Acta Chim. Acad. Sci. Hung.*, 1956, **10**, 281.
11. M. Šišková and E. Erdös, *Coll. Czech. Chem. Comm.*, 1960, **25**, 1729, 3086.
12. D. H. Everett, *Trans. Faraday Soc.*, 1964, **60**, 1803.
13. J. J. Kipling and C. A. J. Langman, unpublished results.
14. M. Šišková and E. Erdös, *Coll. Czech. Chem. Comm.*, 1960, **25**, 2599.
15. F. E. Bartell and H. J. Osterhof, *J. Phys. Chem.*, 1933, **37**, 543.

16. G. A. H. Elton, *J. Chem. Soc.*, 1954, 3813.
17. G. Delmas and D. Patterson, *J. Phys. Chem.*, 1960, **64**, 1827.
18. G. A. H. Elton, *J. Chem. Soc.*, 1952, 1955.
19. G. Schay, L. G. Nagy, and T. Szekrenyesy, *Periodica Polytech.*, 1962, **6**, 91.
20. L. G. Nagy and G. Schay, *Acta Chim. Acad. Sci. Hung.*, 1963, **39**, 365.
21. D. H. Everett, private communication.
22. Y. Fu, R. S. Hansen, and F. E. Bartell, *J. Phys. Chem.*, 1948, **52**, 374.
23. A. Blackburn, J. J. Kipling, and D. A. Tester, *J. Chem. Soc.*, 1957, 2373.
24. G. Schay and L. G. Nagy, *J. Chim. phys.*, 1961, 149.
25. J. C. Eriksson, *Adv. Chem. Phys.*, 1964, **6**, 145.
26. W. M. Jones and W. J. Lewis, private communications.
27. G. J. Young, J. J. Chessick, and F. H. Healey, *J. Phys. Chem.*, 1956, **60**, 394.
28. A. V. Kiselev and L. F. Pavlova, *Bull. Acad. Sci. U.S.S.R.*, 1962, 2030.
29. D. J. Crisp, *J. Colloid Sci.*, 1956, **11**, 356.
30. A. Groszek, *J. Chromatog.*, 1960, **3**, 454.
31. A. V. Kiselev and I. V. Shikalova, *Colloid J. (U.S.S.R.)*, 1962, **24**, 585.
32. S. G. Daniel, *Trans. Faraday Soc.*, 1951, **47**, 1345.

CHAPTER 15

Related Features in the Adsorption of Organic Electrolytes from Solution

TYPES OF ELECTROLYTE

The foregoing chapters are relevant to the adsorption of some, but not of all electrolytes. Three main groups may be recognized: strongly ionized inorganic substances, weakly ionized organic substances, and strongly ionized organic substances.

The adsorption of strongly ionized inorganic compounds is influenced primarily by interactions of the charges carried by relatively small ions. The factors considered in previous chapters are therefore of little relevance. Although this type of adsorption is of considerable importance (e.g. in the stabilization of colloids and in crystal growth), it will not be considered further here. The special case of adsorption by charcoal of electrolytes of all kinds is well reviewed by Steenberg.[1]

The main question in dealing with weakly ionizing inorganic electrolytes is whether the undissociated form is strongly adsorbed, and how adsorption is affected by factors which control the degree of ionization, such as pH. Strongly ionized organic electrolytes often have a small charge associated with a large ion (e.g. detergents, dyestuffs). Adsorption of the organic ion may then be controlled by van der Waals forces operating on the uncharged groups of the ion as well as by electrical forces operating on the charged group. Steenberg suggests that on some charcoals primary adsorption of dyestuff ions occurs mainly through the operation of van der Waals forces, the associated inorganic ion being secondarily adsorbed by electrostatic attraction.

WEAKLY IONIZING ORGANIC SUBSTANCES

The adsorption by charcoal of simple organic solutes which are weakly ionized in aqueous solution (acids, bases, amides), varies with the pH of the solution.[2] For simple aliphatic acids and amines the curve of adsorption against pH follows closely the ionic dissociation curve. Phelps and Peters therefore concluded that adsorption took place through the un-ionized molecules only.[3] Thus if the adsorption isotherm for the free acid or amine is known, the extent of adsorption from any solution of known concentration and pH can be calculated.[4]

The possibility of ion-exchange arises if the solid adsorbent is ionic. Proton transfer can occur in the adsorption of simple peptides from aqueous solution by hydrogen montmorillonite, in addition to physical adsorption. No ion-exchange occurred with sodium and potassium montmorillonites[5] at pH values of about 6.

The nature of the solid surface, as well as that of the solute and solvent, may be altered by change in pH.[6] Ion-exchange may also occur with small quantities of inorganic material present in most charcoals, and with chemically bound complexes (notably acidic groups) present on the surfaces of many charcoals and carbon blacks.

DYESTUFFS AND DETERGENTS

Of the strongly ionizing organic electrolytes, two main groups have received considerable attention. These are ionic dyestuffs and ionic detergents. In both cases, a small charge is associated with a large organic ion. The ion is sufficiently large for some of the factors considered above to be important, but their full effect is modified by electrical interactions, by the presence of the counter-ion, and by susceptibility to added (inorganic) electrolyte. A further complication is the change in properties of the solution which occurs at the critical micelle concentration (c.m.c.). Above this concentration, adsorption characteristics often alter because the concentration of simple ions in solution is reduced, and the concentration of a new species, the micelle, with different interfacial behaviour, grows. Discontinuities in adsorption are sometimes observed at the critical micelle concentration.

Polyelectrolytes constitute a third group. Their adsorption partakes of some of the characteristics of the adsorption of uncharged polymers (Chapter 8) but is also sensitive to the presence of electrolytic impurities and sometimes to pH. These factors are all present in the adsorption of co-polymers of methacrylic acid and vinyl pyridine at the air–water interface.[7] The lowering of surface tension is greatest at the iso-electric point; this is probably mainly due to the effect on solubility, which is lowest at this point.

Adsorption of Dyestuffs

(a) Adsorption at the Liquid–Solid Interface

The adsorption of dyestuffs is mainly of importance at the liquid–solid interface. The phenomenon of dyeing (adsorption by fibres) is discussed by Vickerstaff,[8] in a series of papers by Giles,[9, 10] and elsewhere.[11] In the process of dyeing, a complication sometimes arises which is not normally found with other adsorbents. The fibres may be swollen by the solvent to the extent that they can be penetrated by both solvent and dyestuff.[12] When this occurs, the concept of "surface" cannot be used in the usual way, as the extent of penetration and hence the number of sites for adsorption may vary with the concentration of solution.

In this chapter, attention is given only to adsorption on solids other than textile fibres. Whereas such studies are mainly important as providing models for the dyeing of fibres, the specific case of adsorption by alumina is relevant to the dyeing of anodized aluminium,[13] and that of adsorption by silver halide crystals to the sensitizing of photographic emulsions.

Features which may be important in addition to those discussed in previous chapters include: (a) salt-formation between the dyestuff and the substrate (e.g. some sulphonated compounds with aluminium oxide), (b) ion-exchange at the surface, and (c) formation of chelate complexes, irreversibly, as in lake formation between anthraquinone derivatives (e.g. alizarin) and aluminium oxide,[14] (d) the adsorption of micelles. In the absence of such specific effects, adsorption is subject to the influences described in Chapter 10. In particular, the extent of adsorption may be negligible if the solvent can compete strongly for the solid surface. In some systems two mechanisms operate. The adsorption of basic dyestuffs by clays is in part irreversible (involving the displacement of an equivalent quantity of cation) and in part is reversible physical adsorption. The former process, but not the latter, is independent of concentration.[15]

Strong adsorption of the positively-charged dyestuff ions is assisted by the negative charge which silica surfaces carry when in contact with water. This could be regarded as ion-exchange between the dyestuffs and the hydrogen of surface hydroxyl groups.

Adsorption of micelles must sometimes be postulated because the maximum quantity taken up by the solid is very much greater than corresponds to a monolayer of single dyestuff molecules. Examples are found in the adsorption of basic dyes (e.g. methylene blue, crystal violet, Malachite Green) by silica from aqueous solution.[16] An intermediate case is found in the adsorption of crystal violet and of Orange II by zinc oxide and titanium dioxide, in which a double molecular layer appears to be formed on the surfaces at 25°c. The degree of adsorption decreases markedly, however, with rise in temperature.[17]

Aggregation is also thought to be responsible for an apparent anomaly in the adsorption of dyestuffs. Whereas the extent of adsorption of solutes normally decreases with rise of temperature, that of some dyestuffs increases, e.g. the adsorption of Janus Red B by silica and of Lissamine Green BN by alumina, in each case from water.[18] The adsorption thus appears to be an endothermic process. The normal effect (of decreasing adsorption with rise in temperature) is found if a disaggregating solvent (methanol) is used. Figure 15.1 shows how this effect can be explained in terms of the adsorption of aggregates by the surface. In such experiments, care must be taken to exclude the possibility that apparent high adsorption is due to accelerated fading (e.g. as a result of hydrolysis) of the dyestuff at the higher temperature.

A significant feature of the adsorption of dyestuffs is the presence in a given molecule or ion of several groups which can specifically assist adsorption. Thus the adsorption of a negatively-charged acid dye would be opposed

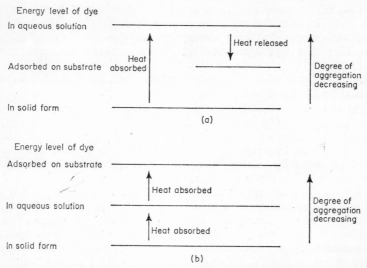

FIG. 15.1. Energy-level diagram for: (a) normal (exothermic) adsorption of dye on a solid surface, (b) anomalous (endothermic) adsorption of a colloidal dye on a solid surface.[18] (Reproduced with permission from the *Journal of the Chemical Society*.)

by the negative charge of a silica surface in contact with water. Orange I, however, is able to form hydrogen-bonds to the surface through its hydroxyl group and is strongly adsorbed.[16] Similarly, such groups may affect orientation at the solid surface. Thus Orange I is held primarily by a covalent bond through the sulphonate group to alumina. At low concentrations, the

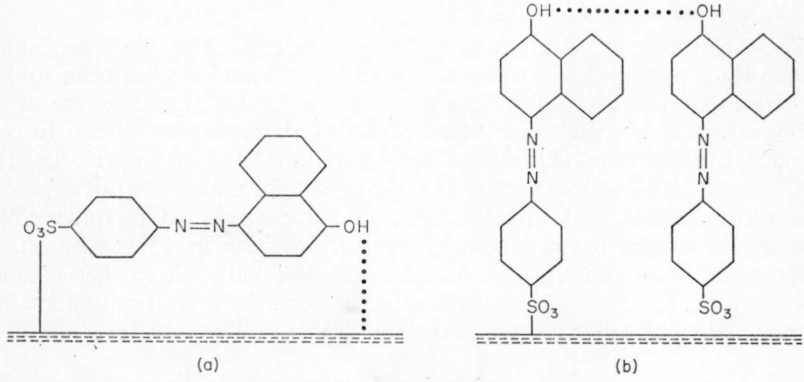

FIG. 15.2. Adsorption of Orange I from aqueous solution by alumina: (a) at low concentrations, (b) at high concentrations.[19] (Reproduced with the permission of Pergamon Press from "Hydrogen Bonding".)

remainder of the molecule lies with the major axis parallel to the surface, because the hydroxyl group can form a hydrogen-bond to the alumina (Fig. 15.2(a)), whereas at high concentrations, this axis is perpendicular to the surface and hydrogen-bonding occurs between adjacent molecules (Fig. 15.2(b)).[19]

A detailed study of the adsorption of dyestuffs by graphite showed that both acid and basic dyes were rapidly adsorbed (only a few minutes being required at room temperature, in contrast to much longer periods required by porous solids). The rate-controlling factor is probably the proportion of molecules or ions which are correctly oriented for adsorption when they strike the surface. In general, a monolayer is readily formed at low concentrations, as judged by a sharp "knee" in the isotherm, but further substantial rises occur at higher concentrations.

At low surface coverages, the molecules are thought to lie flat on the surface, but to re-arrange to a closer packing at high surface coverage, with the ionic groups away from the solid surface. This may give rise to "edge-on" or "end-on" adsorption, depending on the disposition of the charged groups in the molecule. At higher concentrations, multilayer formation or adsorption of micelles occurs.[20] In some cases this gives rise to a second "knee" in the isotherm. This is differentiated from the second "knee" in the adsorption of phenol from water, which is apparently due to re-orientation of the molecule within the monolayer.

The shapes of isotherm can be considered in terms of the classification described in Chapter 7. The type of isotherm depends on the number of aromatic nuclei and of ionic groups in the ion.

A special feature of the adsorption of dyestuffs is that their adsorption on silica gel can be made specific if the silica gel is prepared in the presence of the dyestuff concerned.[21, 22]

(b) Adsorption at the Liquid–Liquid Interface

The adsorption of a small number of dyes (including methylene blue, Congo Red, and Bordeaux extra) at liquid–liquid interfaces has been studied by direct determination of surface excess (see Chapter 2) and by use of the Gibbs equation. In each case one liquid was water; it was placed in conjunction with benzene, chlorobenzene,[23] paraffin,[24] and mercury.[25] The two methods gave widely divergent results, the reason for which was not apparent. Adsorption determined directly rose, in some cases, to a value which remained constant for considerable further increase in concentration. In other cases, it passed through a maximum and fell with further increase in concentration; in one case it became negative. As these results have not been explained, further investigation is desirable.

Ionic Detergents

The behaviour of detergents in contact with textile fibres, like that of dyestuffs, is complex and will not be considered here. The phenomena of detergency as such are described elsewhere.[26, 27] The behaviour of non-ionic detergents in adsorption is covered by previous chapters (e.g. ethylene oxide condensates in Chapter 8).

The adsorption of ionic detergents at all three liquid interfaces has been studied; by comparison with results given in previous chapters, it is possible to deduce in what major respects the presence of ionizing groups affects the adsorption of long-chain compounds. In almost all cases, the results have been obtained for aqueous media. A more detailed treatment, embracing all aspects of the subject, is given in a monograph on colloidal surfactants.[28]

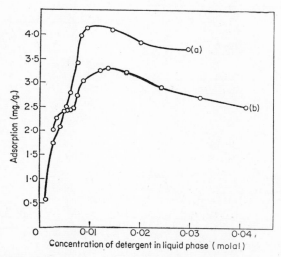

FIG. 15.3. Maxima in isotherms for adsorption[30]: (a) of potassium myristate on graphite at 35°C, (b) of sodium dodecyl sulphate on graphite at 30°C.

(a) *Adsorption at the Liquid–Solid Interface*

The critical micelle concentration (c.m.c.) is important; a break in the adsorption isotherm often occurs at or near this point. Thus, in the adsorption of several sodium alkyl sulphates, quaternary ammonium compounds, and of two non-ionic compounds by carbon black, adsorption reached a limiting value close to the c.m.c.[29] By contrast, the adsorption of sodium dodecyl sulphate and of potassium myristate on graphite was found to pass through a maximum (Fig. 15.3) and then decrease. The latter effect was attributed to repulsion between ionic "head groups" in the adsorbed layer,[30] though it is difficult to understand how the maximum value was reached against such repulsion. This effect is even more apparent in the adsorption of sodium dodecylbenzene sulphonate by cotton from 0·01N HCl. It may possibly be related to the formation of increasingly large micelles as the concentration increases.[31] An alternative suggestion is that a sudden formation of micelles occurs on the surface from the single ions already adsorbed. Desorption of single ions then ceases and is replaced by desorption of micelles, with consequent alteration in the dynamic equilibrium.[32]

Adsorption also reaches a maximum at the c.m.c. in adsorption of cationic and anionic detergents by polar solids.[33, 34] The isotherm is S-shaped (Fig. 15.4) at low concentrations. This probably corresponds to electrostatic adsorption of one layer, followed by adsorption of a second layer by van der Waals attraction. It is also possible, but less likely, that a phase-change occurs, the

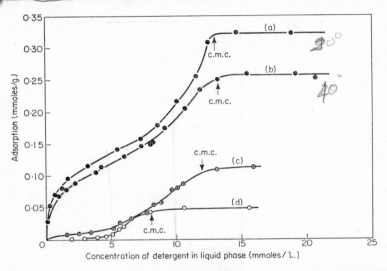

FIG. 15.4. Adsorption of detergents from aqueous solution by alumina[33]: (a) dodecyl-ammonium chloride at 20°C, (b) dodecylammonium chloride at 40°C, (c) dodecylpyridinium bromide at 20°C, (d) sodium dodecyl sulphate at 20°C. (Reproduced with the permission of Butterworths from "Proceedings of the Second International Congress of Surface Activity", Vol. III, 1957.)

long-chain ions originally being arranged with the major axis parallel to the surface, but subsequently perpendicular to it. In the latter orientation, repulsion between the ions would account for lower adsorption than would be expected for a corresponding neutral molecule. For Graphon, a solid more adequately characterized, the limiting adsorption of sodium dodecyl-benzene sulphonate corresponded to a molecular area of about 65 sq. Å, which must imply considerable repulsion between adjacent molecules.[35]

There is, however, considerable evidence that double layers can be formed on solids. Schulman has given convincing evidence that sodium dodecyl sulphate is adsorbed in double layers on barium sulphate from solutions of relatively high concentration.[36] The first layer is assumed to be chemisorbed (a phenomenon which would not occur with Graphon) and the second to be physically adsorbed with the ionized group oriented towards the aqueous phase. This arrangement is compatible with the re-wetting of the particles at high detergent concentration. It is also claimed that sodium laurate is adsorbed in double layers on barium sulphate.[37]

The difference in behaviour on different solids may arise from degree of polarity of the surface. Koganovskii has pointed out that micelles of amphipathic materials form in water with the ionic groups on the outside.[38] As these groups are not strongly attracted to a non-polar surface such as graphite, he considers the adsorption of a hemi-spheroidal micelle which would allow the hydrocarbon chains to be in contact with (and parallel to) the solid surface. On a polar surface, however, adsorption would involve the ionic groups of the amphiphile, and the complete micelles would be adsorbed. Thus the limiting adsorption of a given substance, per unit area, would be twice as great on polar as on non-polar surfaces.

Koganovskii finds a sharp rise in the adsorption of hexadecyl and octadecyl sulphates from aqueous solutions on graphite at concentrations well above the c.m.c., but close to the solubility limit. This is attributed to some form of re-organization of the micellar structure.

(b) Adsorption at the Liquid–Vapour Interface

In the earlier literature, considerable attention was paid to curves of surface tension against concentration which passed through a pronounced minimum. These curves are now believed to be unreliable.[39–41] The minima may have been due to: (a) the presence of impurities in the solute; as little as 0·05% of dodecanol in sodium dodecyl sulphonate produces a very marked minimum in the curves for aqueous solution, and a smaller effect is produced by sodium dodecyl sulphate,[42,43] (b) the presence of heavy metals in the water used, (c) the use of a method in which true equilibrium was not reached.[44]

The extent of adsorption, as for non-electrolytes at this interface, is most often calculated from surface tension data by use of the Gibbs equation (see Chapter 11). There has been considerable discussion as to whether this equation should be modified when applied to electrolytes, because each molecule can dissociate into two or more ions. Most detergents are uni-univalent electrolytes, and the equation for the surface excess is sometimes used with the inclusion of the factor 2:

$$\Gamma_2^{(1)} = -\frac{1}{2RT}\frac{d\gamma}{d\ln a_2}.$$ (15.1)

More generally (cf. equation 11.1)

$$-\frac{d\gamma}{RT} = \frac{d\pi}{RT} = \sum \Gamma_i . d\ln a_i,$$ (15.2)

where Γ_i is the surface concentration of component i.

If conditions of electrical neutrality are introduced, it can be shown[45, 46] for dilute solutions of an electrolyte A^+B^-, where B^- is the surface-active ion, that

$$\frac{d\pi}{RT} = \Gamma_B^{(1)} d \ln a_{A^+} . a_{B^-} + (\Gamma_{A^+}^{(1)} - \Gamma_{B^-}^{(1)}) d \ln a_{A^+} . a_{OH^-}, \qquad (15.3)$$

where a_{A^+} and a_{B^-} are the respective ionic activities in the bulk phase. If activities are now replaced by molalities with associated activity coefficients, then as

$$m_{A^+} = m_{B^-} = m_{AB},$$

so that

$$\ln a_{A^+} . a_{B^-} = 2 \ln m_{AB} . f_{A^+, B^-}$$

(where f_{A^+, B^-} is the mean activity coefficient), it follows that equation (15.3) becomes:

$$\frac{d\pi}{RT} = 2\Gamma_B^{(1)} . d \ln m_{AB} f_{A^+, B^-} + (\Gamma_{A^+}^{(1)} - \Gamma_{B^-}^{(1)}) d \ln m_{A^+} . m_{OH^-} f_{A^+, OH^-}. \quad (15.4)$$

If m_{OH^-} is constant, and also $\Gamma_{A^+}^{(1)} = \Gamma_{B^-}^{(1)}$, this becomes

$$\frac{d\pi}{RT} = 2\Gamma_B^{(1)} . d \ln m_{AB} . f_{A^+, B^-}, \qquad (15.5)$$

which is equivalent to equation (15.1) with B^- as component 2.

On the other hand, if $\Gamma_{A^+}^{(1)}$ tends to zero,

$$-\frac{d\gamma}{RT} = \Gamma_B^{(1)} . d \ln m_{AB} . f_{A^+, B^-} \qquad (15.6)$$

and the factor 2 is eliminated. It can also be shown that in the presence of a constant excess of a neutral salt having the common ion A^+,

$$-\frac{d\gamma}{RT} = \Gamma_B^{(1)} . d \ln m_{B^-} f_{B^-} \qquad (15.7)$$

and, if further assumptions are made, that in the presence of a constant excess of another salt (say $D^+ E^-$),

$$-\frac{d\gamma}{RT} = \Gamma_B^{(1)} . d \ln m_{B^-}. \qquad (15.8)$$

For the long-chain sodium alkyl sulphates (decyl, dodecyl, tetradecyl, and hexadecyl), adsorption tends to a limiting value at the c.m.c., and at that point corresponds approximately to the value given by equation (15.1).[47, 48]

In the presence of excess of sodium chloride, the adsorption of sodium dodecyl sulphate is in accordance with equation (15.7)[47] and in the presence of other salts (not containing a common ion), with equation (15.8).[49] In terms of the simple form of the Gibbs equation, the factor 2 is required in the first case, but not in the other two.

An isotherm obtained by use of the radioactive technique is shown in Fig. 15.5 for adsorption of sodium di-n-octyl sulphosuccinate at the water–air interface.[50] The value of the adsorption at the "knee" corresponds to the Γ calculated from the Gibbs equation without the factor 2, and then gives a molecular area for the adsorbate which appears reasonable by reference to molecular models. The authors suggested that the absence of the factor 2 might be explained by assuming that the adsorbate was the free acid, not the sodium salt. There is supporting evidence for this pheno-

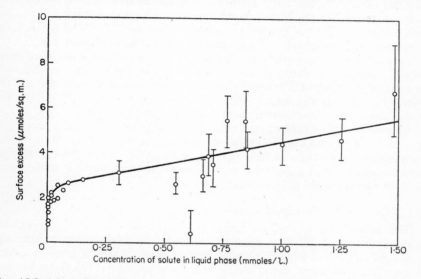

FIG. 15.5. Adsorption of sodium di-n-octyl sulphosuccinate at the water–air interface.[50] (Reproduced with permission from the *Proceedings of The Royal Society*.)

menon[51] which has also been referred to[52] as ion-exchange between H_3O^+ and Na^+. On the other hand, the effect may be due to ionic impurities in the water.[53] The rise in the isotherm at higher concentrations might be due to multilayer formation, but could be a spurious effect caused by the radio-activity of increasingly concentrated solution immediately below the adsorbed monolayer.

If equation (15.1) is applied to surface tension data for solutions of sodium dodecyl sulphate in water, it gives molecular areas of 26·3 and 18·1 sq. Å at 25°c and 15°c respectively.[44] The latter value has been criticized as unrealistically small. Measurements on the tritiated material give a value[54, 55] of 33 sq. Å at 25°c.

Equation (15.7) has been used for solutions of sodium undecylsulphonate, and gave a value of 52 sq. Å per ion, compared with a mean value of 51 sq. Å per ion obtained directly from foam experiments.[56]

By measuring rates of desorption from the surface film, Davies has shown that the Gibbs equation without the factor 2 applies to the adsorption of cetyl trimethylammonium bromide at the interface between water (0·01 N in hydrochloric acid) and high-boiling petroleum ether.[57] The hydrochloric acid was in large excess, as the concentration of the detergent was in the region of 10^{-4} to 10^{-5} N. The choice of a quaternary ammonium salt eliminates the possibility of adsorption of an undissociated molecule.

Thus, in general, the factor of 2 is required for the more concentrated solutions (either because the detergent is adsorbed in a hydrolysed form or because the gegen-ion is as strongly adsorbed as the detergent ion), but not for dilute solutions (as the gegen-ion is only weakly adsorbed), nor for solutions containing excess of a neutral salt which has a common ion.

In adsorption of potassium laurate at the water–air interface, there exists a substantial range of concentration below the critical micelle concentration over which adsorption appears to remain constant although the surface tension continues to fall with rise in concentration.[58] Above the c.m.c. the surface tension remains constant.

Measurements of surface concentrations using the radio-active technique are relevant to the consideration of the structures of soap films.[59] The technique has also been useful in studies of competitive adsorption involving two surface-active agents.[60] To a first approximation, surface activity was found to be inversely proportional to the c.m.c.

(c) Adsorption at the Liquid–Liquid Interface

The adsorption of sodium lauryl (dodecyl) sulphate at the n-decane–water interface at 20°C has been measured by the emulsion technique and has been calculated from interfacial tension data. In the absence of added electrolyte, molecular areas calculated from equation (15.5) agree well with those measured by the emulsion technique (44·2 and 48·8 sq. Å, respectively at the c.m.c.).[61] Other studies have confirmed the use of equation (15.5) when added electrolyte is absent and of equation (15.7) in the presence of added electrolyte, e.g. careful measurements by Kling and Lange of alternate sodium alkyl sulphates, from C_8 to C_{20}, at the water–heptane interface.[62] Quite small amounts of electrolyte may be sufficient, however, to alter the behaviour of surface films so that the results agree with equation (15.7) particularly at low concentrations of detergent.[53] Impurities present in added salts may accumulate slowly and produce anomalous results.[63]

The relevance of considering long-chain electrolytes together with the corresponding polar but non-ionic compounds is shown in Fig. 15.6. There is clearly only a small difference in the free energy of adsorption of alkyl sulphates and of the corresponding alcohols at the oil–water interface; the slight difference in slope of the two curves is due to a difference in temperature of the two measurements. The greatest contribution to the free

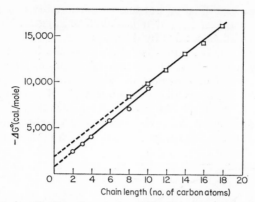

FIG. 15.6. Free energies of adsorption at the aliphatic hydrocarbon oil–water interface[64] as a function of chain-length for n-alkyl sulphates at 50°C (□), and n-alkyl alcohols at 20°C (○). (Reproduced from the "Proceedings of the Third International Congress of Surface Activity", Verlag der Universitätsdruckerei, Mainz.

energy of adsorption comes from the hydrocarbon chain (820 cal./mole for each CH_2 group at 20°C; 800 cal./mole at 50°C). The contribution of the polar group is small by comparison, with relatively small differences between the ionic and non-ionic groups (—OH, 800 cal./mole at 20°C; —COOH, 1630 cal./mole at 20°C; —NMe_2^+, 950 cal./mole at 20°C; —HSO_4^-, 1800 cal./mole at 50°C).[64]

Equations of State

An equation of state for the adsorbed monolayer can be obtained by combining the equation for the adsorption isotherm with the appropriate form of the Gibbs equation. The term for the free energy of adsorption included in the former differs according to whether or not the adsorbate is ionized. For ionized adsorbates, it must include the work done against electrical forces, which is additional to work done in adsorption of a neutral molecule. The consequence for equations of state has been discussed by a number of authors.[63,65-67]

KINETICS OF ADSORPTION

The kinetics of adsorption of organic electrolytes is more complicated than is the case for the non-electrolytes considered in Chapter 13. Important features are the effects of impurities, the formation of micelles, and the formation of an electrical double layer. Impurities in particular may produce changes over long periods of time.

At low concentrations of ionic detergents, the surface excess is proportional to the square root of time for the sodium alkyl sulphates in the initial stages of forming the surface layer. This would be in accordance with the

control of the rate by diffusion, but the proportionality constants are incon-
sistent with this conclusion, and it is therefore suggested that there is an
energy barrier as well as control by diffusion.[68]

For another system, the aqueous solutions of the dyestuff Benzpurpurin
10 B, it is suggested that, as adsorption takes place, an electrical double
layer is formed which complicates the simple kinetic expression given in
equation (13.7).[69] This would provide one type of energy barrier. In the
presence of considerable quantities of inorganic electrolyte, the rate of
adsorption of long-chain ionic detergents seems to depend, however, on
diffusion of the organic ion to the surface, not on electrical barriers.

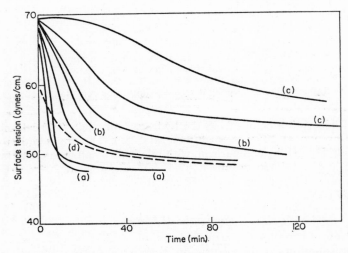

FIG. 15.7. Effect of impurities on the surface tension of an aqueous solution of sodium
hexadecyl sulphate (5×10^{-5} M) at 40°C: (a) singly distilled water, (b) doubly distilled
water, (c) trebly distilled water,[73] (d) conductivity water.[72] (Reproduced from the
Australian Journal of Chemistry.)

For simple aqueous solutions of sodium stearate, $\log \Gamma$ is a linear function
of $\log t$, as would be required if the rate-controlling process were diffusion
to a sub-surface of the kind envisaged by Ward and Tordai (Chapter 13);
similar results are obtained with solutions 10^{-4} to 10^{-3} M in sodium
hydroxide.[70]

At high relative concentrations of ionic detergents, equilibrium is reached
quickly if micelles are present.[71] This may be due to repulsion between the
single ions being adsorbed and the highly charged micelles, though the
adsorption of neutral micelles has also been considered.[72]

In the later stages of adsorption, the surface tension of some solutions
may decrease slowly over a period of hours. One effect of impurities is
shown for dilute solutions of an ionic detergent in Fig. 15.7. For more

concentrated solutions, the effect is much less marked. Inorganic impurities have a much greater effect than do non-ionic material (e.g. hexadecanol) or homologues of the detergent. Sutherland concluded that the change in surface tension was due to slow diffusion of the impurities to the surface, the process being correspondingly slower the more dilute the solution with respect to impurities.[73] Other explanations of the slow change have been suggested. For fatty acid soaps, it is possible that hydrolysis (especially in the presence of carbon dioxide) causes slow release of the free acid which is more surface-active than the soap. The slow evaporation of rather involatile impurities (e.g. dodecanol present in alkyl sulphonates) may be important.[74]

FROTH FLOTATION

This method of separating a mineral from the unwanted parts of its ore (the gangue) depends on the attachment of one of the components (usually the mineral) in finely divided form to a stream of air-bubbles rising through the water. Adsorption from solution is relevant to the process because "collectors" are added to the aqueous phase to increase the wettability of one of the components. Increasing the hydrophobic character of the solid particle increases the likelihood of its becoming attached to an air bubble, whereas a hydrophobic solid is likely to remain in the bulk of the aqueous phase. Surface-active agents ("frothers") are also used to stabilize the froth or foam.

The use of collectors depends on the adsorption by the solid of an organic compound containing a polar group. The latter provides the point of attachment to the surface; the organic residue confers the required hydrophobic properties. Examples are organic acids, xanthates, amines, alkyl phosphates and sulphates. Selectivity is discussed by Schulman and Smith in terms of pH of the solution and size of the metal ion in the surface of the mineral.[79] Collectors have been classified according to their behaviour in adsorption.[80]

The effect of pH is very marked in many cases and the results suggest that the effective process is adsorption of an ion (an anion for basic minerals such as metallic oxides, a cation for acidic minerals such as quartz, either anion or cation for salt-like minerals such as barytes). The neutral molecule is not often effective as a collector, but this does not imply that it is not adsorbed at any point of the pH range.

Flotation may occur readily when a monolayer or much less than a monolayer of collector has been adsorbed, but is sometimes completely inhibited by the adsorption of a second layer. In such systems the ions in the second layer have the opposite orientation to that in the first layer, i.e. with the polar groups towards the solution; this would make the particle hydrophilic once more.

The use of collectors may be assisted by the "activators" which promote the adsorption of the collector on the material to be floated, and by "depressants" which reduce adsorption of the collector on the material which is to remain in the aqueous phase. Depressants compete for adsorption sites on the solid surface but do not contain hydrophobic groups (e.g. sodium sulphide). The relevant adsorption phenomena are often ionic in character if the mineral is ionic and so fall outside the scope of this book. Some physical adsorption of neutral molecules probably occurs. Adsorption of neutral molecules may be effective if the mineral is not ionic in character, e.g. phenolic compounds are used effectively in flotation of coal.

REFERENCES

1. B. Steenberg, "Adsorption and Exchange of Ions on Activated Charcoal", Almquist and Wicksells, Uppsala, 1944.
2. A. H. Andersen, *Acta Pharmacol. Toxicol.*, 1947, **3**, 199.
3. H. J. Phelps and R. A. Peters, *Proc. Roy. Soc.*, 1929, A, **124**, 554.
4. J. J. Kipling, *J. Chem. Soc.*, 1948, 1483.
5. D. J. Greenland, R. H. Laby, and J. P. Quirk, *Trans. Faraday Soc.*, 1962, **58**, 829.
6. G. Lopatin and F. R. Eirich, "Proceedings of the Third International Congress of Surface Activity", Verlag der Universitätsdruckerei, Mainz, 1960, Vol. II, p. 97.
7. I. R. Miller and A. Katchalsky, "Proceedings of the Second International Congress of Surface Activity", Butterworths, London, 1957, Vol. I, p. 159.
8. T. Vickerstaff, "The Physical Chemistry of Dyeing", Oliver and Boyd, London and Edinburgh, 1954, 2nd edition.
9. C. H. Giles, "Compte-rendu du 31ᵉ Congrès Internationale de Chimie Industrielle", Liege, 1958.
10. C. H. Giles, *Textile Res. J.*, 1961, **31**, 141.
11. "The Physical Chemistry of Dyeing and Tanning", *Discuss. Faraday Soc.*, **16**, 1954.
12. H. R. Chipalkatti, C. H. Giles, and D. G. M. Vallance, *J. Chem. Soc.*, 1954, 4375.
13. C. H. Giles, "Proceedings of the Conference on Anodising", Paper 13, 1961.
14. C. H. Giles, H. V. Mehta, C. E. Stewart, and R. V. R. Subramanian, *J. Chem. Soc.*, 1954, 4360.
15. P. H. Plesch and R. H. S. Robertson, *Nature*, 1948, **161**, 1020.
16. M. M. Allingham, J. M. Cullen, C. H. Giles, S. K. Jain, and J. S. Woods, *J. Appl. Chem.*, 1958, **8**, 108.
17. W. W. Ewing and F. W. J. Liu, *J. Colloid Sci.*, 1953, **8**, 204.
18. C. H. Giles, J. J. Greczek, and S. N. Nakhwa, *J. Chem. Soc.*, 1961, 93.
19. C. H. Giles in "Hydrogen Bonding", (D. Hadzi, ed.), Pergamon Press, London, 1959, p. 449.
20. J. W. Galbraith, C. H. Giles, A. G. Halliday, A. S. A. Hassan, D. C. McAllister, N. Macaulay, and N. W. MacMillan, *J. Appl. Chem.*, 1958, **8**, 416.
21. F. H. Dickey, *Proc. Nat. Acad. Sci., U.S.A.*, 1949, **35**, 227.
22. R. G. Haldeman and P. H. Emmett, *J. Phys. Chem.*, 1955, **59**, 1039.
23. C. W. Gibby and C. C. Addison, *J. Chem. Soc.*, 1936, 119.
24. C. W. Gibby and C. C. Addison, *J. Chem. Soc.*, 1936, 1306.
25. C. W. Gibby and C. Argument, *J. Chem. Soc.*, 1940, 596.

26. J. L. Moilliett, B. Collie, and W. Black, "Surface Activity", Spon, London, 1961, 2nd edition.
27. "Surface Activity and Detergency", (K. Durham, ed.), MacMillan, London, 1961.
28. K. Shinoda, T. Nakagawa, B. I. Tamamushi, and T. Isemura, "Colloidal Surfactants", Academic Press, New York and London, 1963, Chap. 3.
29. G. R. F. Rose, A. S. Weatherburn, and C. H. Bayley, *Textile Res. J.*, 1951, **21**, 427.
30. M. L. Corrin, E. L. Lind, A. Roginsky, and W. D. Harkins, *J. Colloid Sci.*, 1949, **4**, 485.
31. A. Fava and H. Eyring, *J. Phys. Chem.*, 1956, **60**, 890.
32. R. D. Vold and N. H. Swaramakrishnan, *J. Phys. Chem.*, 1958, **62**, 984.
33. B. Tamamushi and K. Tamaki, "Proceedings of the Second International Congress of Surface Activity", Butterworths, London, 1957, Vol. III, p. 449.
34. B. Tamamushi and K. Tamaki, *Trans. Faraday Soc.*, 1959, **55**, 1007.
35. A. C. Zettlemoyer, C. H. Schneider, and J. D. Skewis, "Proceedings of the Second International Congress of Surface Activity", Butterworths, London, 1957, Vol. III, p. 472.
36. B. D. Cuming and J. H. Schulman, *Austral. J. Chem.*, 1959, **12**, 413.
37. N. A. Held and K. N. Samochwalov, *Kolloid Z.*, 1935, **72**, 13.
38. A. A. Koganovskii, *Colloid J. (U.S.S.R.)*, 1962, **24**, 597.
39. G. D. Miles and L. Shedlovsky, *J. Phys. Chem.*, 1944, **48**, 57.
40. A. P. Brady, *J. Phys. Chem.*, 1949, **53**, 56.
41. A. Lake, A. S. C. Lawrence, and O. S. Mills, "Proceedings of the Second International Congress of Surface Activity", Butterworths, London, 1957, Vol. I, p. 200.
42. L. Shedlovsky, J. Ross, and C. W. Jakob, *J. Colloid Sci.*, 1949, **4**, 25.
43. R. Ruyssen, *Bull. Soc. chim. belges*, 1953, **62**, 97.
44. E. J. Clayfield and J. B. Matthews, "Proceedings of the Second International Congress of Surface Activity", Butterworths, London, 1957, Vol. I, p. 172.
45. B. A. Pethica, *Trans. Faraday Soc.*, 1954, **50**, 413.
46. E. A. Guggenheim, "Thermodynamics", North Holland Publishing Co., Amsterdam, 1949.
47. H. Lange, *Kolloid Z.*, 1957, **153**, 155.
48. R. Matuura, H. Kimizuka, S. Miyamoto, and R. Shimozawa, *Bull. Chem. Soc. Japan*, 1958, **31**, 532.
49. R. Matuura, H. Kimizuka, and K. Yatsunami, *Bull. Chem. Soc. Japan*, 1959, **32**, 646.
50. D. J. Salley, A. J. Weith, A. A. Argyle, and J. K. Dixon, *Proc. Roy. Soc.*, 1950, A, **203**, 42.
51. C. M. Judson, A. A. Lerew, J. K. Dixon, and D. J. Salley, *J. Chem. Phys.*, 1952, **20**, 519.
52. J. W. James and B. A. Pethica, "Proceedings of the Third International Congress of Surface Activity", Verlag der Universitätsdruckerei, Mainz, 1960, Vol. I, p. 227.
53. D. A. Haydon and J. N. Phillips, *Trans. Faraday Soc.*, 1958, **54**, 698.
54. G. Nilsson, "Proceedings of the Second International Congress of Surface Activity", Butterworths, London, 1957, Vol. I, p. 141.
55. G. Nilsson, *J. Phys. Chem.*, 1957, **61**, 1135.

56. F. van Voorst Vader and M. van den Tempel, "Proceedings of the Third International Congress of Surface Activity", Verlag der Universitätsdruckerei, Mainz, 1960, Vol. I, p. 248.
57. J. T. Davies, *Trans. Faraday Soc.*, 1952, **48**, 1052.
58. C. P. Roe and P. D. Brass, *J. Amer. Chem. Soc.*, 1954, **76**, 4703.
59. J. M. Corkhill, J. F. Goodman, D. R. Haisman, and S. P. Harrold, *Trans. Faraday Soc.*, 1961, **57**, 821.
60. K. Shinoda and K. Kinoshita, *J. Colloid Sci.*, 1963, **18**, 174.
61. E. G. Cockbain, *Trans. Faraday Soc.*, 1954, **50**, 874.
62. W. Kling and H. Lange, "Proceedings of the Second International Congress of Surface Activity", Butterworths, London, 1957, Vol. I, p. 295.
63. D. A. Haydon and F. H. Taylor, *Trans. Faraday Soc.*, 1962, **58**, 1233.
64. D. A. Haydon and F. H. Taylor, "Proceedings of the Third International Congress of Surface Activity", Verlag der Universitätsdruckerei, Mainz, 1960, Vol. I, p. 157.
65. D. A. Haydon and F. H. Taylor, *Phil. Trans.*, 1960, A, **252**, 225.
66. D. A. Haydon and F. H. Taylor, *Phil. Trans.*, 1960, A, **252**, 255.
67. J. T. Davies and E. K. Rideal, "Interfacial Phenomena", Academic Press, London, 1961.
68. R. Matuura, H. Kimizuka, S. Miyamoto, R. Shimozowa, and K. Yatsunami, *Bull. Chem. Soc. Japan*, 1959, **32**, 404.
69. K. S. G. Doss, *Kolloid Z.*, 1938, **84**, 138; 1939, **86**, 206.
70. S. N. Flengas and E. K. Rideal, *Trans. Faraday Soc.*, 1959, **55**, 339.
71. N. K. Adam and H. L. Shute, *Trans. Faraday Soc.*, 1938, **34**, 758.
72. G. C. Nutting, F. A. Long, and W. D. Harkins, *J. Amer. Chem. Soc.*, 1940, **62**, 1496.
73. K. L. Sutherland, *Austral. J. Chem.*, 1959, **12**, 1.
74. L. F. Evans, quoted by K. L. Sutherland, *Rev. Pure Appl. Chem. (Australia)*, 1951, **1**, 35.
75. V. I. Klassen and V. A. Mokrousov, "An Introduction to the Theory of Flotation", (translated by J. Leja and G. W. Poling), Butterworths, London, 1963.
76. "Recent Developments in Mineral Dressing", The Institution of Mining and Metallurgy, London, 1953.
77. K. L. Sutherland and I. W. Wark, "Principles of Flotation", Australasian Institute of Mining and Metallurgy (Inc.), Melbourne, 1955.
78. R. B. Booth. *In* "Foams", Reinhold Publishing Corp., New York, 1953, p. 243.
79. J. H. Schulman and T. D. Smith, "Recent Developments in Mineral Dressing", The Institution of Mining and Metallurgy, London, 1953, p. 393; cf. German translation in *Kolloid Z.*, 1952, **126**, 20.
80. J. Leja, "Proceedings of the Second International Congress of Surface Activity", Butterworths, London, 1957, Vol. III, p. 273.

CHAPTER 16

The Use of Columns in Adsorption

GENERAL

Adsorption from solution is often carried out as a "batch" process, i.e. a solid and solution are agitated together until equilibrium is established. Most of the discussion in previous chapters has been based on this procedure. In the practical use of adsorption, however, two advantages may accrue from allowing the solution to flow through a column of adsorbent. The first is that continuous operation may be maintained. The second is that a column can be very effective in the separation of a number of solutes present in the same solution or in the complete removal of a single solute from its solvent. Any preferential adsorption shown by the adsorbent can be exploited in a column which gives the effect of the summation of the separations achieved in many consecutive batch processes. The last feature is particularly important in the separation of substances which are very similar, e.g. homologues and geometrical isomers, and in the fractionation of polymers.[1,2]

The specific topic of chromatography is considered briefly below. Besides separation at the liquid–solid interface, separations (e.g. of synthetic surface-active agents) have been achieved at the liquid–liquid and the liquid–air interfaces, in the latter case by using columns of foam.[3,4]

MATHEMATICAL TREATMENT

The mathematical treatment of adsorption by a column has been approached from two slightly different points of view. In the "theory of chromatography" (outlined below) the primary emphasis is on the separation in terms of the components of the solutions. In a second treatment, greater emphasis is given to the characteristics of the adsorbent which forms the column.

The second treatment is derived from equations used to describe the passage of a gas through a column of granular adsorbent, for which several treatments have been given. Hinshelwood's equation[5] has been modified specifically to deal with adsorption from solution. It deals with conditions of initial "breakdown" of the column, i.e. the first appearance of detectable traces of the adsorbate in the stream leaving the column. The "breakdown time", τ, of a column of length λ is given by:

$$\tau = \frac{1}{kc_0} \left[\ln\left(e^{kN_0\lambda/L} - 1\right) - \ln\left(c_0/c' - 1\right) \right], \qquad (16.1)$$

where c_0 is the concentration of the adsorbate entering the column, c' is the concentration of the adsorbate leaving the column, L is the linear flow-rate of the solution, N_0 is the number of active centres of unit activity/c.c. of adsorbent, and k is a constant, except in conditions specified below.

An exact treatment of flow through a column leads to intractable equations. The above equation is therefore based on approximations; those relevant to the adsorption of solutes are: (i) that the rate of adsorption is fast relative to the rate of percolation, (ii) that the adsorbate is strongly held, so that desorption can be neglected under the conditions used.

In practice $e^{kN_0\lambda/L}$ is much greater than 1, so we can re-write equation (16.1) as:

$$\tau = \frac{1}{kc_0}\left[\frac{kN_0\lambda}{L} - \ln(c_0/c'-1)\right]. \tag{16.2}$$

Hence it can be deduced that:

$$\tau + \tau_0 = N_0\lambda/c_0L, \tag{16.3}$$

where

$$\tau_0 = [\ln(c_0/c')-1]/kc_0, \tag{16.4}$$

i.e. that breakdown time is a linear function of column length.

In time τ, a "breakdown volume", Y, of solution has flowed through the column; Y is given by:

$$Y = F\tau = AL\tau, \tag{16.5}$$

where F is the total flow-rate of the solution and A is the cross-sectional area of the column. As rates of diffusion are lower in solution than in the vapour phase, and the equation may be applied to adsorption by porous solids, k must be replaced by $k_D.L^{\frac{1}{2}}$. It can then be shown that

$$Y = Y_0 - KF^{\frac{1}{2}}, \tag{16.6}$$

where

$$Y_0 = \frac{AN_0\lambda}{c_0} \quad \text{and} \quad K = \frac{A^{\frac{1}{2}}\ln(c_0/c'-1)}{k_Dc_0},$$

i.e. the breakdown volume is a linear function of the square root of the total flow rate.

A column is formed of granules of adsorbent which may vary in size. Under practical conditions, N_0 does not alter with granule diameter, but for porous adsorbents, the accessibility of the sites of adsorption may be affected. Accessibility may be represented by the ratio:

$$\frac{\text{external surface of granule}}{\text{volume of granule}} = \frac{\pi d^2}{1/6\pi d^3} = \frac{6}{d},$$

where d is the diameter of the granule. We replace k by k^*/d, and equation (16.1) reduces to the form

$$\tau = \tau_0' - k_1 d \tag{16.7}$$

where

$$\tau_0' = \frac{N_0\lambda}{c_0L} \quad \text{and} \quad k_1 = \frac{\ln(c_0/c'-1)}{k^*c_0}.$$

In experiments with 0.1 N aqueous solutions of acetic acid and columns of charcoal, it was shown that equations (16.3) and (16.6) were obeyed exactly and equation (16.7) approximately.[6] The effect of initial concentration was as given by equation (16.1), i.e. the curve of τ against $1/c_0$ was linear when c_0 was small (i.e. when $\ln(c_0/c'-1)$ was much less than $kN_0\lambda/L$) and fell off towards the c/c_0 axis as c_0 increased. A different effect of particle diameter is found if adsorption is diffusion controlled.[7]

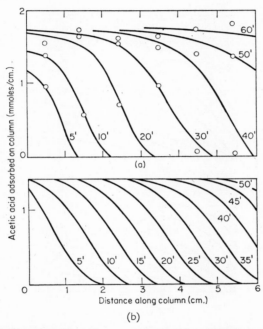

FIG. 16.1. Distribution of acetic acid on a charcoal column in adsorption from aqueous solution as a function of time: (a) experimental, (b) calculated. See note in text.[6] (Reproduced with permission from the *Journal of the Chemical Society*.)

For moderate flow-rates, the distribution of solute on the column can be calculated with considerable accuracy by assuming that a condition close to equilibrium is maintained. Distributions can be calculated for both adsorption and desorption. For this purpose the column is considered in terms of a series of small elements and the equilibrium between solution and adsorbate for each element is calculated for successive small intervals of time, in terms of mass transfer through the column. This is an adaptation of a method first devised for calculating the distribution of adsorbed gas on columns.[8] Typical results for adsorption and desorption are shown in Figs. 16.1 and 16.2.

The experimental results have not been corrected for the quantity of solution held mechanically by the column in interstices, but not adsorbed. This accounts for the difference in height of the curves. Apart from this, the shapes of the calculated curves follow those drawn through the experimental points very closely, showing that distribution on the column under these conditions is governed predominantly by the shape of the adsorption isotherm.

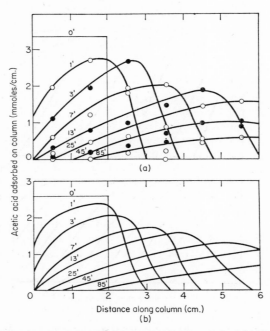

Fig. 16.2. Distribution of acetic acid during desorption, by water, from a column of charcoal as a function of time: (a) experimental, (b) calculated. See note in text.[6] (Reproduced with permission from the *Journal of the Chemical Society*.)

A very different situation is found in the industrial dechlorination of water by passing it through beds of charcoal. This involves much more dilute solutions and much higher flow rates. Moreover, although the process initially involves adsorption (of hypochlorous acid), this is followed by reaction at the surface (cf. Chapter 6).

A slmplified version of the treatment given above leads to the equation

$$\ln \frac{c_0}{c'} = \frac{k}{L^{\frac{1}{2}}} \tag{16.8}$$

if the total process is diffusion-controlled. If, however, the process is controlled by a slow reaction at the surface, the equation

$$\ln \frac{c_0}{c'} = \frac{k}{L} \qquad (16.9)$$

should apply. The latter is found to apply to the dechlorination process.[9] The constant k was found not to be inversely proportional to d as suggested above, but was given by the equation

$$k = \frac{a}{d} - b, \qquad (16.10)$$

where a and b are constants.

CHROMATOGRAPHY

Detailed accounts of chromatography are given elsewhere in a number of monographs and discussions.[10,11] This section is concerned only with the application to chromatography of ideas developed earlier in this book, and is thus limited to adsorption chromatography at the liquid–solid interface. It is arguable that individual use of chromatography is more empirical than it need be because not enough attention is paid to the chemical nature of the solid comprising the column, its porosity, the nature of the solvent, the shape of the relevant adsorption isotherms, or even the possibility that one or more component of the solution or the eluting agent may be chemisorbed. Some authors have, however, emphasized the relevance of these factors in the use of chromatography.[12] The free energy of adsorption may be a guide to the order of breakdown times for a given column to the solutes in a particular solution.[13]

After the formation of the initial chromatogram, development may be continued to the stage at which the material originally adsorbed passes into the eluate. Three procedures can then be distinguished:

  (i) frontal analysis, in which the original solution is passed continuously through the column,
 (ii) elution development, in which the solutes, once adsorbed, are removed by a pure solvent,
(iii) displacement development, in which the solutes, once adsorbed, are displaced by passing through the column a solution of a very strongly adsorbed displacing agent.

The accounts given in Chapters 4, 7, and 10 should be relevant in deciding the appropriate choice of solvent in elution development and of displacing agent in displacement development. The adsorption isotherms are relevant, as the least strongly adsorbed solute travels most rapidly through the column.[14] Extremely strong adsorption should be avoided in the separation of similar substances, as this reduces discrimination. Thus if fatty acids of differing degrees of unsaturation are to be separated on a polar adsorbent (e.g. silica gel), it is useful to esterify the acids before separation.[15] In this way differences between the hydrocarbon chains become more important relative to the common carboxyl group which, if free, might determine the nature of the adsorption.

The theory of frontal analysis, due to Tiselius, has been described by Claesson.[16] In the simple form of displacement development, a stationary state is reached in which the boundaries of all the solutes move down the column at the same rate as the front of the displacing agent. Then, providing that displacement is quantitative,

$$\frac{f_1(c_1)}{c_1} = \frac{f_2(c_2)}{c_2} = \ldots = \frac{f_d(c_d)}{c_d}, \tag{16.11}$$

where the subscript d refers to the displacing agent. Thus if the concentration of the displacing agent is the same in a set of experiments, the concentration of a given displaced substance in the eluate is always the same and can be used as a means of identification.

Theory of Chromatography

A full theoretical treatment of chromatography would have to take into account at least three factors: (a) the relative positions on a column taken up by the several solutes from the original solution, this being determined by the respective free energies of adsorption, (b) the rates of passage of the solutes down the column, which depend on the respective rates of desorption,[17] (c) the characteristics of the column.

The theoretical treatment of chromatography is, in part, analogous to that of adsorption on a column of adsorbent as described above. It is usually based on the assumption that the rate of flow of solution through the column is slow relative to the rate of attainment of equilibrium. The equation for the adsorption isotherm can then be used to relate the extent of adsorption to the concentration of solution at any point in the column. This approach leads to a first-order partial differential equation to describe the situation at any point in the column. If, however, adsorption is diffusion controlled, a second-order equation is obtained, as Amundson has pointed out.[18] The latter type of equation can only be solved analytically in very special cases which Amundson discusses.

In addition to the use of columns for chromatography, discs which involve radial flow have also been considered in both of the above respects.[19]

Adsorption Involving One Solute

The simplest treatment is for a solution containing only one solute.[20-22] The requirement that mass is conserved when a solution flows through the column and solute is transferred from solution to adsorbent leads to the equation:

$$\frac{\partial c}{\partial x} + [\alpha + M f'(c)] \frac{\partial c}{\partial V} = 0, \tag{16.12}$$

where c is the concentration of the solution at a distance x from the top of the column, α is the interstitial volume per unit length of the column,† V is the volume of solution which has entered the column at a given time (which is also the volume of solution which has passed any point in the column up to that time if the column was initially filled with solution). The equation for the adsorption isotherm is

$$Q = Mf(c), \tag{16.13}$$

where Q is the amount of solute adsorbed per unit length of column, and M is the weight of adsorbent per unit length. The function $f(c)$ has to be specified in any use of the treatment; f' is the first derivative of f. The integrated form of equation (16.13) is

$$c = \phi(V - x[\alpha + Mf'(c)]), \tag{16.14}$$

where ϕ is an arbitrary constant. To meet the appropriate boundary conditions (for $V = 0$ and $x = 0$), ϕ must be a discontinuous function of x.

The behaviour of any adsorbate on the column depends markedly on the shape of the adsorption isotherm. An equation of the Freundlich or Langmuir form has often been used in conjunction with equation (16.14). This is probably satisfactory for dilute solutions, but as no account is taken of the adsorption of solvent, the accuracy of the treatment necessarily falls in any consideration of more concentrated solutions.

If the adsorption isotherm is linear, a band is formed on the column with a sharp leading edge and a sharp trailing edge. If the isotherm is concave to the concentration axis, there is a sharp leading edge and a diffuse trailing edge, with the converse situation if the isotherm is convex. (These situations can be described in terms of a Freundlich equation:

$$Q = Mc^{1/n}. \tag{16.15}$$

They correspond to $\frac{1}{n} = 1$, $\frac{1}{n} < 1$, and $\frac{1}{n} > 1$, respectively.) The second case occurs most commonly in practice. The effect of a sigmoid isotherm has also been considered.[24]

An equation for the isotherm has also been based on the assumption that adsorption is governed by a simple Law of Mass Action expression,[25] but this is probably more readily applied to ion-exchange processes than to adsorption based on van der Waals forces. For the purposes of chemical engineering, it has been developed by Vermeulen,[26-29] on the basis of work by Thomas.[30]

The nature of the isotherm to be used in the above treatment is not usually specified. Smith considers that this should be the isotherm of preferential adsorption (the "composite" isotherm discussed in Chapter 3), not the individual isotherm.[31] For most systems of practical significance, the distinction is of little importance because very dilute solutions are used

† Allowance can also be made for the volume occupied by adsorbed solute.[23]

(cf. equation 3.9). It has been shown to be important, however, in dealing with pairs of completely miscible liquids.[32] A more detailed discussion is given by Baylé and Klinkenberg.[33,34]

Adsorption of Several Solutes

When adsorption occurs from a solution containing i solutes, equation (16.13) is replaced by i equations, each having the general form:

$$Q_i = Mf_i(c_1, c_2, c_3, ..., c_i) \qquad (16.16)$$

and a strict treatment would then be based on a knowledge of the effect on the adsorption of any one solute of the presence of all others. As indicated in Chapter 9, little direct experimental evidence is available about this, and this limits the application of the theory. The simplest situation would be independent adsorption of each solute. As this is thought unlikely to occur in practice, Wilson has suggested that equations of the form

$$Q_i = MA \frac{a_i c_i}{1 + \sum a_i c_i} \qquad (16.17)$$

should be used, where A is a constant characteristic of the adsorbent and a_i a constant characteristic of the ith solute.

Glückauf illustrates the development of a chromatogram containing two substances on the basis of this type of equation; he considers development both with a pure solvent and with a solution containing a displacing solute.[35] Complete separation of the solutes cannot always be obtained if both isotherms are described by Freundlich equations.[36] Further mathematical treatments are given by Glückauf[37] and Weiss.[38]

Chromatography of an aqueous solution containing a mixture of fatty acids suggested that the above treatment, based on the use of Freundlich equations, was satisfactory, but that different constants were needed in adsorption from the mixture from those which applied to adsorption of each solute separately.[39] It has, however, been suggested that some degree of ion-exchange at the surface of the charcoal used as the adsorbent might complicate the mathematical treatment.[40]

REFERENCES

1. L. A. Baker and R. J. P. Williams, *J. Chem. Soc.*, 1956, 2352.
2. M. J. R. Cantow, R. S. Porter, and J. F. Johnson, *J. Polymer Sci.*, 1963, C, **1**, 187.
3. L. Shedlovsky, *Ann. N.Y. Acad. Sci.*, 1948, **49**, 279.
4. K. Shinoda and K. Machio, *J. Phys. Chem.*, 1960, **64**, 54.
5. C. J. Danby, J. G. Davoud, D. H. Everett, C. N. Hinshelwood, and R. M. Lodge, *J. Chem. Soc.*, 1946, 918.
6. J. J. Kipling, *J. Chem. Soc.*, 1948, 1487.
7. D. H. Sharp, Ph.D. Thesis, University of London, 1950.

8. R. F. Barrow, C. J. Danby, J. G. Davoud, C. N. Hinshelwood, and L. A. K. Staveley, *J. Chem. Soc.*, 1947, 401.
9. V. Magee, Ph.D. Thesis, University of London, 1955.
10. "Chromatographic Analysis", *Discuss. Faraday Soc.*, 1949, Vol. 7.
11. "Chromatography", (E. Heftmann, ed.), Reinhold, New York, 1961.
12. H. G. Cassidy, "Adsorption and Chromatography", Interscience, New York, 1951, 1st edition.
13. T. M. Rovinskaya, *Colloid J.* (*U.S.S.R.*), 1962, **24**, 184.
14. L. Hagdahl. *In* "Chromatography", (E. Heftmann, ed.), Reinhold, New York, 1961, p. 59.
15. L. J. Morris. *In* "Chromatography", (E. Heftmann, ed.), Reinhold, New York, 1961, p. 430.
16. S. Claesson, *Discuss. Faraday Soc.*, 1949, **7**, 34.
17. C. H. Giles. *In* "Chromatography", (E. Heftmann, ed.), Reinhold, New York, 1961, p. 33.
18. N. R. Amundson, *J. Phys. Chem.*, 1950, **54**, 812.
19. L. Lapidus and N. R. Amundson, *J. Phys. Chem.*, 1950, **54**, 821.
20. J. N. Wilson, *J. Amer. Chem. Soc.*, 1940, **62**, 1583.
21. J. Weiss, *J. Chem. Soc.*, 1943, 297.
22. D. DeVault, *J. Amer. Chem. Soc.*, 1943, **65**, 532.
23. W. M. Smit, *Discuss. Faraday Soc.*, 1949, **7**, 38.
24. E. Glückauf, *J. Chem. Soc.*, 1947, 1302.
25. J. E. Walter, *J. Chem. Phys.*, 1945, **13**, 229.
26. N. K. Heister and T. Vermeulen, *Chem. Eng. Progr.*, 1952, **48**, 505.
27. T. Vermeulen and N. K. Hiester, *Ind. Eng. Chem.*, 1952, **44**, 636.
28. T. Vermeulen and N. K. Hiester, *J. Chem. Phys.*, 1954, **22**, 96.
29. T. Vermeulen and N. K. Hiester, *Chem. Eng. Progr.*, 1958, **55**, Symposium Series, No. 24, p. 61.
30. H. C. Thomas, *J. Amer. Chem. Soc.*, 1944, **66**, 1664.
31. W. R. Smit and A. van den Hoek, *Rec. Trav. chim.*, 1957, **76**, 561.
32. A. van den Hoek and W. R. Smit, *Rec. Trav. chim.*, 1957, **76**, 577.
33. G. C. Baylé and A. Klinkenberg, *Rec. Trav. chim.*, 1957, **76**, 593.
34. A. Klinkenberg and G. C. Baylé, *Rec. Trav. chim.*, 1957, **76**, 607.
35. E. Glückauf, *Proc. Roy. Soc.*, 1946, A, **186**, 35.
36. E. Glückauf, *J. Chem. Soc.*, 1947, 1321.
37. E. Glückauf, *Discuss. Faraday Soc.*, 1949, **7**, 12.
38. C. Offord and J. Weiss, *Discuss. Faraday Soc.*, 1949, **7**, 26.
39. F. H. M. Nestler and H. G. Cassidy, *J. Amer. Chem. Soc.*, 1950, **72**, 680.
40. E. Glückauf, *J. Amer. Chem. Soc.*, 1951, **73**, 849.

CHAPTER 17

The Use of Adsorption from Solution in Measurement of Surface Area

INTRODUCTION

The measurement of the specific surface areas of solids is becoming increasingly important, and a wide range of methods is now available.[1] Adsorption is useful for solids of relatively high specific surface area. Methods based on physical adsorption of gases (nitrogen, argon, krypton) at low temperatures (together with electron microscopy for non-porous solids) are as near to being standard methods as we have. At one time they were regarded as having a sound theoretical basis in the B.E.T. equation, but more recently this assumption has been criticized.[2]

It can at least be claimed that the B.E.T. equation provides a standard procedure for determining empirically the point at which monolayer adsorption is complete on non-porous solids. Moreover, interest both in the adsorption of gases by solids and in the determination of surface areas is such that this method is likely to remain of primary importance. Simplified experimental procedures are being developed and the theoretical background is being continually examined.

In this situation, the use of adsorption from solution in determining surface areas is of secondary importance. It should be considered, however, for two reasons. First, it was (relatively) more widely used before methods based on gas adsorption had been extensively developed. Secondly, it still has the attraction that the experimental procedure is much simpler than in any method requiring vacuum apparatus, and, if routine measurements on a large number of samples are involved, it is usually much quicker.

For a method based on adsorption from solution to be sound, two fundamental requirements should be fulfilled. It should be possible: (i) to determine the conditions under which a complete monolayer of a given component is formed on the solid (or alternatively the composition of a mixed monolayer), (ii) to assign an accurate value to the area occupied by the adsorbed molecular species; this normally means that the orientation of the adsorbed molecules must be known with confidence.

In the light of the points discussed in earlier chapters, it will be seen that there is no general equation which can be relied upon to give the monolayer capacity of a solid from any given set of measurements. The adsorption of sparingly soluble solutes gives the greatest promise, but adsorption from a given solution may lead to the formation of a complete monolayer on one solid, but not on another, even at the highest attainable concentrations; similarly a complete monolayer may be formed on a particular solid from one solvent but not from another. Moreover, the fact that the observed data can be fitted by a particular equation (e.g. one of the Langmuir form) does not mean that the constants of the equation are necessarily related to the surface area of a monolayer.

The value of Schay's analysis of isotherms which have a section corresponding to a region of almost constant composition in the adsorbed phase has been considered in Chapter 4. As yet, however, not enough is known for a satisfactory prediction to be made as to whether the isotherm for a given system (solution and solid) will be of the form suitable for Schay's analysis.

Measurements of surface area have usually been considered in terms of physical adsorption, but chemisorption does not have to be excluded from consideration automatically. In determining the monolayer capacity when chemisorption occurs it is important to ensure that interaction with the solid does not go beyond the surface to a continuous reaction with the bulk of the solid. Further, if chemisorption is followed by physical adsorption, it must be possible to establish what proportion of the adsorbate is held in each layer.

The effective molecular area of the adsorbate normally remains constant for a range of solids if the molecules are physically adsorbed in a close-packed layer. A major change in molecular area may occur if the orientation changes between one group of solids and another. If adsorption occurs at specific sites on the surface, e.g. through hydrogen-bonding or in chemisorption, the effective molecular area is a function primarily of the spacing of the sites, and so can vary from one solid to another. If only a proportion of the sites can be occupied even at the maximum degree of adsorption (see Chapter 7), the effective molecular area of the adsorbed molecules may vary even more from one solid to another as this proportion changes.

For volatile molecules, an effective molecular area can be obtained from adsorption of the vapour on an appropriate solid of known specific surface area. This gives the most likely value for the molecular area adopted in adsorption from solution, though it is not certain that the same orientation of the molecules occurs in both cases. If this basis of establishing the molecular area has to be used, the method is clearly valuable only if a large number of measurements of adsorption from solution are to be carried out. The solid used in adsorption of the vapour should be chemically as similar as possible to the solids on which adsorption from solution is to be carried out.

For non-volatile adsorbates, the area of the molecule can usually be calculated from tables of bond lengths and bond angles. The effective molecular area in the adsorbed layer depends, however, on the orientation of the molecules and (in the case of molecules of irregular shape) on the packing which is adopted. The latter is often suggested by the arrangement of the molecules in the corresponding crystal. The combined effect of packing and orientation can only be assessed, however, by reference to adsorption on a solid of known specific surface area.

It follows that methods based on adsorption from solution have only an empirical basis. For any sets of results to be regarded as reliable, the method should first have been "calibrated" by reference to a well-charac-terized solid† having the same surface characteristics as the samples being investigated. The accuracy of the results is liable to vary according to the system being examined and for some may not be high. Experimental error is usually smaller, and often very much smaller, than uncertainties involved in finding the correct monolayer value and choosing the correct molecular area for the adsorbate. As an extreme example, the two main orientations

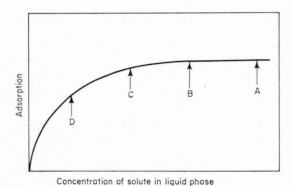

Fig. 17.1. Choice of initial concentration of solution for single-point determination of surface area.[38]

of stearic acid at a surface give molecular areas differing by a factor of five. Yet in many investigations it appears that the only molecular area which received consideration was 20·5 sq. Å, the value given in close-packed films on aqueous substrates, not even on a solid adsorbent.

Similarly, in the adsorption of methylene blue from aqueous solution, a molecular area has repeatedly been quoted from Paneth[3] without a proper assessment of its significance. Paneth did obtain this value by calibration, but by a procedure which now seems to have been unnecessarily crude.[4]

Experimentally, many results have been based on one-point determinations. The appropriate monolayer value can only be obtained by considering the shape of the whole adsorption isotherm. When this has been done, it may be possible to use single determinations subsequently, though only if the concentration of solution is properly chosen. It must be at a point (A) on the isotherm (Fig. 17.1) such that after adsorption has taken place, the

† This usually involves, at least, measurement of the specific surface area by adsorption of nitrogen or krypton.

final concentration still lies within the region of approximately constant adsorption (B). It would not be satisfactory to start with an initial concentration at C, leading to final concentration at D; not only would adsorption not correspond to completion of a monolayer, but its extent would vary with the specific surface area of the sample and of the ratio of weights of sample and solution.

ADSORPTION FROM SOLUTIONS OF SOLIDS

The use of these systems has often been unduly empirical, the points outlined above being ignored. The results are then necessarily suspect, except for comparative purposes when different samples of the same substance were used.

When substantial sections of the isotherm have been obtained, the results have sometimes been expressed in terms of the Freundlich equation, though this (cf. Chapter 7) has little significance. As the isotherms for adsorption from solutions of solids usually tend towards a limiting value, the limit has often been taken to correspond to the coverage of the surface with a complete monolayer of the solute. Alternatively, the Langmuir equation with c/c_0 replacing p, has been used to determine the limiting value. In Chapter 7 it was shown that although adsorption comes to a very definite limit, this does not always correspond to the formation of a complete monolayer of solute on the surface. Moreover, the orientations (and consequently the molecular areas) assumed for the adsorbed molecules have sometimes been shown not to be justified. The method is therefore suspect unless the behaviour of a given solute–solvent system has been explored for the type of adsorbent being examined.

Fatty Acids

In this method, adsorption of long-chain fatty acids from organic solutions seems to have been used frequently because it has been thought that molecular areas are known accurately from adsorption of the acids on aqueous substrates.[5] Authors have used the molecular area appropriate to close-packed films at high compression, apparently without reflecting that a different orientation is adopted at lower surface pressures. For stearic acid, the two molecular areas differ by a factor of about five. It is therefore important to consider what orientation is adopted on solid surfaces, as examples of each orientation are now known.

In one of the early proposals for the use of fatty acids, Harkins and Gans (cf. Chapter 7) established that it was valid to assume perpendicular orientation for titania powders, but their method has often been followed uncritically. Comparison with areas established by low-temperature nitrogen adsorption has similarly shown its use for determining areas of some forms of alumina[6] and of some metals.[7, 8]

For the specific case of metals, Tingle has produced three categories.[9]

In the first (e.g. cadmium, copper, zinc) the metal is normally covered with a thick permeable layer of oxide which reacts with fatty acids sufficiently to make surface area determinations imprecise. In the second category are metals (e.g. aluminium, chromium, nickel) which form compact oxide films on which only a monolayer of acid is formed, possibly by chemisorption. In the third category are metals (e.g. platinum, gold, silver) which form a thin oxide film or none at all. For the second and third categories, satisfactory surface area determinations can normally be made. The kind of agreement found is between 0·210 sq. m./g. for an iron powder from adsorption of stearic acid, and 0·173 sq. m./g. calculated from the mean diameter of particles, assumed to be spherical.[10]

When adsorbates such as fatty acids are chemisorbed, the molecular area should not necessarily be assumed to be the same as in a close-packed physically adsorbed film. It is more likely to be determined by the spacings of the crystal lattice of the substrate (cf. Chapter 7), and so to vary considerably between different solids. Unless the appropriate molecular area is known, the value obtained for the specific surface area should not be expected to agree closely with those obtained by other methods. Extensive chemical reaction has, rather surprisingly, been found between perfluorooctanoic acid and glass.[11]

The results quoted in Chapter 7 show that on some solids, the acid molecule is adsorbed with its major axis parallel to the surface and that from some solvents, a complete monolayer of acid is not adsorbed. This factor is probably responsible for the conflicting reports on the measurement of surface areas of clays. Orr and Bankston claim that the method is suitable,[12] but this has been disputed by Birrell who finds large discrepancies.[13] He suggests that this may be related to the nature of the surface of the clays, a layer of amorphous alumina helping to give concordance. In view of subsequent investigation of the adsorption of stearic acid by alumina[14] this seems likely, but further evidence is needed before it can be regarded as proved. An approximately five-fold discrepancy between nitrogen and stearic acid areas of sand and silica[15] suggests that in this case the stearic acid was probably adsorbed with the molecules parallel to the surface, as has been found for silica,[14] and not perpendicular to it as was assumed in calculating the areas.

The use of radioactive stearic acid (labelled with ^{14}C) enables measurements to be made on solids, having the appropriate type of surface,[16, 17] down to specific surface areas of less than 1 sq. m./g.

The method breaks down if the adsorbent has very narrow pores. Thus for a particular silica gel used as a catalyst support, low-temperature nitrogen adsorption gave 809 sq. m./g. against 22 sq. m./g. from adsorption of stearic acid.[18] For a number of other catalyst supports, much closer agreement was found, though it is questionable whether the assumption should have been made in all cases that the stearic acid molecules were adsorbed perpendicular to the surface. There is evidence that stearic acid will penetrate into pores with openings of about 8 Å in diameter, but not into those of 5 Å in diameter.[19]

An attempt has been made to apply the modified form of the B.E.T equation (Chapter 5) to adsorption of stearic acid on calcium carbonate. The specific surface areas obtained from the supposed monolayer values agreed with those obtained by other physical methods when benzene was used as the solvent.[20] When carbon tetrachloride was the solvent, however, the calculated specific surface areas dropped to between one-half and two-thirds of the previous values, although the experimental points were still well fitted by the equation.[21]

Esters

Ewing investigated the adsorption of esters from solution in benzene by zinc oxide powders.[22] He showed that twice as many molecules of methyl stearate as of glycol dipalmitate were adsorbed per unit weight, which implies that the chains of these molecules were oriented perpendicular to the surface. There seems no reason, however, why this particular type of substrate should be chosen for surface area determination.

Phenols

Adsorption of phenol from aqueous solution has been recommended for determining surface areas. A single curve is obtained if adsorption per unit area is plotted for three carbon blacks of different surface characteristics, the surface areas having been obtained by a different method. Comparison of any other isotherm (with adsorption expressed in terms of unit weight of adsorbent) with this curve enables the specific surface of the adsorbent to be calculated.[23]

As in the adsorption of stearic acid, an attempt has been made to apply the modified B.E.T. equation to adsorption of phenol. With carbon tetrachloride as the solvent and clays and silica gels as the adsorbents, this gives "monolayer values" from which specific surface areas were calculated in close agreement with those obtained from low-temperature adsorption of nitrogen.[24] It would be interesting to know whether this agreement is fortuitous and whether it would be obtained if different solvents were used. In an investigation on porous carbons it was shown that the ratio of the specific surface areas determined from adsorption of phenol and of nitrogen or argon varied widely between samples.[25] The "phenol area" was always the lower, by a factor between 2·5 and 5. This might be due to a molecular sieve effect. Thus, if the method could be satisfactorily established, the results might give useful information about the micro-pore structure of the adsorbent.

Iodine

Adsorption of iodine has been used with organic solvents and with aqueous solutions containing various concentrations of potassium iodide. The method has serious limitations, as is implied by the discussion of adsorption of iodine in Chapter 7. On some solids (e.g. silica), iodine is only weakly adsorbed. For glass, a specific surface area calculated from adsorption from solution in carbon tetrachloride was less than 40% of that estimated microscopically,[26] and the latter, for reasons of surface roughness, is likely to have been low.

For solids generally, the appropriate molecular area to use is uncertain. Thus although the limiting adsorption of iodine from carbon tetrachloride on active magnesia was proportional to the specific surface area deduced from low-temperature adsorption of nitrogen,[27] the use of a molecular area of 21·2 sq. Å for iodine gave an area 35% lower than that calculated from adsorption of nitrogen. The low value may well result from adsorption on regular sites spaced so as to give less than close-packing, thus requiring a greater effective area per molecule. The solids used were slightly porous, but partial exclusion of iodine seems unlikely, as the extent of adsorption was proportional to that of nitrogen for a range of samples.

On other solids, e.g. carbon black, adsorption has been shown to vary according to the organic solvent used (see Chapter 7) and to the degree of homogeneity of the surface. Under standardized conditions, however, reliable results can be obtained for adsorption from aqueous potassium iodide. The carbon blacks must, however, be of low volatile content or have been "de-volatilized", i.e. heated to a sufficiently high temperature to remove oxygen complexes from the surface.[28] The latter process is likely to alter (usually to reduce) the specific surface area. If devolatilization is not carried out, the results are likely to be high,[29] probably, in the main, because the solution reacts slowly with the carbon black.

Diphenyl Guanidine

The importance of surface complexes is evident in the adsorption of diphenyl guanidine from 0·01 M solution in benzene. Two very different straight lines are obtained correlating specific surface area and adsorption of diphenyl guanidine, depending on whether the surface is covered by or is free from oxygen complexes.[28] The reason for this divergence has not been fully elucidated.

Dyestuffs

The use of dyestuffs is particularly attractive because colorimetric analysis of the solutions is readily carried out. This can be useful, not only in the conventional measurements, but also in the use of chromatographic methods for determining monolayer values.[30] This advantage, however, is offset by several limitations:

(i) Most dyestuffs consist of large planar molecules, the three dimensions of which differ considerably, e.g. the length, breadth, and thickness of the methylene blue ion (Fig. 17.2) are 16·0 Å, 8·4 Å, and 4·7 Å, respectively. Thus three major orientations at an interface

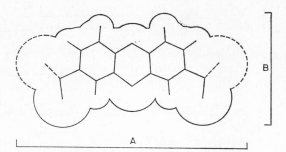

FIG. 17.2. Dimensions of the cation of methylene blue.[4] (Reproduced with the permission of the Society of the Chemical Industry from the *Journal of Applied Chemistry*.)

are possible with widely differing molecular areas. For a few dyestuffs, the orientation has been shown to differ according to the substrate. One cyanine dye has been shown to occupy a molecular area on silver chloride more than twice that which it occupied on silver iodide.[31]

(ii) Most dyestuffs are ionic, and the adsorption is subject to a number of extra factors, such as the pH of the solution and the presence of other electrolytes (see Chapter 15).

(iii) Dyestuffs may occur in solution in the form of micelles, which in some systems are adsorbed in preference to single molecules.[32]

An example of the variations found in adsorption on one class of solids is given by values of the molecular area of methylene blue when adsorbed on carbons. Paneth's original use of this dyestuff was based on a value of 1 sq. m./mg. adsorbed. This appears to have been calculated from a formula applicable to spherical molecules. (The calibration against crushed diamond of surface area established microscopically appears rather crude in view of what is now known of the molecular roughness of most surfaces.) This corresponds to a value of 54 sq. Å per molecule if the dyestuff used was anhydrous. Graham found a molecular area of 197 sq. Å for adsorption on Graphon[33] compared with later values of 108 and 102 sq. Å for graphite and Spheron 6, respectively.[4] Boehm and Hoffman found a value of either 78 or 138 sq. Å, depending on whether the carbon had been graphitized or not.[34] It is also clear that the extent to which dyestuffs are adsorbed by a given solid depends on the solvent used.[35]

An Ideal Adsorbate

With these limitations in mind, Giles has defined the characteristics which would be required for a solute to give reliable surface area determinations.[36] It should:

(i) be highly polar, to ensure adsorption by polar solids;
(ii) have hydrophobic properties, to permit adsorption by non-polar solids;
(iii) have a small molecule, preferably planar and likely to be adsorbed in a close-packed monolayer with the major axis perpendicular to the surface, so that estimation of the molecular area can be made accurately;
(iv) should not be highly surface-active, as micelle-formation at the surface is undesirable;
(v) should be coloured, for ease of analysis;
(vi) should be readily soluble in water, for convenience in use, but also soluble in non-polar solvents so that it can be used with water-soluble solids.

No single substance meets all these requirements, but Giles has met with considerable success in using p-nitrophenol.[37] Water is the preferred solvent, but hydrocarbons (benzene, xylene, and heptane) have also been used. For non-porous solids, good correlation has been obtained with results for adsorption of nitrogen. For solids believed to be porous, the value given by adsorption of p-nitrophenol is sometimes lower, suggesting the presence of pores which admit nitrogen but are too small to admit p-nitrophenol. In the case of some organic pigments, better penetration appears to be achieved by p-nitrophenol.

Adsorption of p-nitrophenol on finely-divided metals is much greater than that of nitrogen. This may be due to aggregation of the particles in the static experiments used for adsorption of gases, but efficient separation when shaken with a solution. If this is so, the point deserves much wider consideration. The adsorption of p-nitrophenol from solution may prove to be suitable for determining the surface areas of fibres.

ADSORPTION OF COMPLETELY MISCIBLE LIQUIDS

In view of what is now known about adsorption from completely miscible liquids, they would generally be regarded as unsuitable for the present purpose. In particular, the composite isotherm gives no indication of the monolayer value of either component. The use of dilute solutions is deluding, as has been shown[38] by an analysis of results for adsorption of the lower fatty acids from dilute aqueous solutions by charcoals.

It has been claimed, however, that one type of isotherm can be used in the determination of specific surface areas.[39] This is the isotherm which has a substantial linear section. The significance of such isotherms has been discussed in Chapter 4. They would be useful when they could be used for deducing the composition of a mixed monolayer. It is then necessary, of course, to know the molecular area of each component in order to calculate the specific surface area. Probably the most serious practical limitation in the use of this method is that it would rarely be possible to predict which liquid system would give the required type of isotherm with a given solid. Once this were established, however, the same system could be used repeatedly for further samples of the same kind of solid.

Similarly, equation (4.13) might be used to find the total number of molecules in the adsorbed layer of unit weight of solid. As this equation applies only to systems in which the two components have the same molecular area and which show ideal behaviour in both the liquid and the adsorbed phase, it is unlikely to be widely applicable. In particular, it will not generally be known in advance of carrying out an experiment whether or not the last condition is obeyed.

THE SIGNIFICANCE OF SPECIFIC SURFACE AREAS

In any measurement of the specific surface area of a solid, the value of a particular determination depends on the use which is to be made of it. This is mainly important when porous solids are being considered. Thus the area obtained depends on the size of the "measuring" molecule. This should clearly be small if an estimate of the maximum available surface area is required. If, however, the area available to a large molecule (e.g. that of a coloured impurity or of a polymeric material) is important, this maximum area may be quite misleading, and a much larger "measuring" molecule should be used.

This consideration leads to the possibility that the distribution of pore sizes could be estimated by using a series of solutes of varying molecular sizes as "molecular probes", giving the surface area associated with pores above a critical size.[40] Dubinin and Zaverina outlined this approach in examining a set of charcoals. It has not been seriously reconsidered, however, since the difficulties in using this method for determining surface areas were pointed out.

The method is likely to be of value almost exclusively for examining micropores, as the range of size of suitable solutes is severely limited. When more is known of the effective size of polymers in solution and of their behaviour with porous solids, it might be possible to extend the method to larger pores.

REFERENCES

1. C. Orr and J. M. DallaValle, "Fine Particle Measurement", The MacMillan Co., New York, 1959.
2. D. M. Young and A. D. Crowell, "Physical Adsorption of Gases", Butterworths, London, 1962.
3. F. Paneth and A. Radu, *Ber.*, 1924, B **57**, 1221.
4. J. J. Kipling and R. B. Wilson, *J. Appl. Chem.*, 1960, **10**, 109.
5. N. K. Adam, "The Physics and Chemistry of Surfaces", Oxford University Press, London, 1941, 3rd edition.
6. A. S. Russell and C. N. Cochran, *Ind. Eng. Chem.*, 1950, **42**, 1332.
7. C. Orr, H. G. Blacker, and S. L. Craig, *J. Metals*, 1952, **4**, 657.

8. H. A. Smith and J. F. Fuzek, *J. Amer. Chem. Soc.*, 1946, **68**, 229.
9. E. D. Tingle, *Trans. Faraday Soc.*, 1950, **46**, 93.
10. E. B. Greenhill, *Trans. Faraday Soc.*, 1949, **45**, 625.
11. J. W. Shepherd and J. P. Ryan, *J. Phys. Chem.*, 1959, **63**, 1729.
12. C. Orr and P. T. Bankston, *J. Amer. Ceram. Soc.*, 1952, **35**, 58.
13. K. S. Birrell, *New Zealand J. Sci. Technol.*, 1957, B, **38**, 588.
14. J. J. Kipling and E. H. M. Wright, *J. Chem. Soc.*, 1964, 3535.
15. A. G. Loomis, *Producers Monthly*, 1954, **28**, No. 11, 19.
16. M. C. Kordecki and M. B. Gandy, *Internat. J. Appl. Radiation Isotopes*, 1961, **12**, 27.
17. H. P. Dibbs, P. R. Gorla, and C. M. Lapointe, "Proceedings of the Third International Congress of Surface Activity", Verlag der Universitätsdruckerei, Mainz, 1960, Vol. III, p. 242.
18. H. E. Ries, M. F. L. Johnson, and J. S. Melik, *J. Phys. Chem.*, 1949, **53**, 638.
19. J. J. Kipling and E. H. M. Wright, unpublished results.
20. E. Suito, M. Arakawa, and T. Arakawa, *Bull. Inst. Chem. Res., Kyoto Univ.*, 1955, **33**, 1.
21. E. Suito, M. Arakawa, and T. Arakawa, *Bull. Inst. Chem. Res., Kyoto Univ.*, 1955, **33**, 8.
22. W. W. Ewing, *J. Amer. Chem. Soc.*, 1939, **61**, 1317.
23. A. V. Kiselev and I. V. Shikalova, *Zavodskaya Lab.*, 1958, **9**, 1074.
24. H. P. Boehm and W. Gromes, *Angew. Chem.*, 1959, **71**, 65.
25. W. Naucke, *Brennstoff-Chem.*, 1963, **44**, 302.
26. A. J. Urbanic and V. R. Damerell, *J. Phys. Chem.*, 1941, **45**, 1245.
27. A. C. Zettlemoyer and W. C. Walker, *Ind. Eng. Chem.*, 1947, **39**, 69.
28. W. R. Smith, F. S. Thornhill, and R. I. Bray, *Ind. Eng. Chem.*, 1941, **33**, 1303.
29. P. Besson and J. Sanlaville, *J. Chim. phys.*, 1950, **47**, 108.
30. H. Benesi, *Anal. Chem.*, 1960, **32**, 1410.
31. J. F. Padday, *Trans. Faraday Soc.*, 1964, **60**, 1323.
32. M. M. Allingham, J. M. Cullen, C. H. Giles, S. K. Jain, and J. S. Woods *J. Appl. Chem.*, 1958, **8**, 108.
33. D. Graham, *J. Phys. Chem.*, 1955, **59**, 896.
34. A. Clauss, H. P. Boehm, and U. Hofmann, *Z. anorg. Chem.*, 1957, **290**, 35.
35. J. W. Galbraith, C. H. Giles, A. G. Halliday, A. S. A. Hassan, D. C. McAllister, N. Macaulay, and N. W. MacMillan, *J. Appl. Chem.*, 1958, **8**, 416.
36. C. H. Giles, T. H. MacEwen, S. N. Nakhwa, and D. Smith, *J. Chem. Soc.*, 1960, 3973.
37. C. H. Giles and S. N. Nakhwa, *J. Appl. Chem.*, 1962, **12**, 266.
38. J. J. Kipling in "Porous Carbons", (R. L. Bond, ed.), Academic Press, London, 1966.
39. G. Schay, L. G. Nagy, and T. Szekrényesy, *Periodica Polytech.*, 1960, **4**, 95.
40. M. Dubinin and E. D. Zaverina, *Acta Physicochim. U.R.S.S.*, 1936, **4**, 647.

AUTHOR INDEX

Numbers in parentheses are reference numbers and are included to assist in locating references where the author's name is not mentioned in the text.

Numbers in italics refer to the page on which the reference is listed.

Abe, R., 155 (58, 59), 156 (58), *159*
Abram, J. C., 166 (18), *189*
Adam, N. K., 9 (30), *20*, 41 (20), *68*, 115 (44), *132*, 192 (2), 197 (2), 198 (2), 201 (2), *216*, 274 (71), *278*, 291 (5), *297*
Adamson, A. W., 9 (34), *20*, 192 (3), *216*
Addison, C. C., 10 (38 44), *20*, 214 (53), *217*, 236 (28), 238 (40–42), 239 (40), 240 (43–47), 241 (48), *243*, 266 (23, 24), *276*
Akamatu, H., 16 (69), *21*
Alexander, A. E., 237 (29), *243*
Allen, K. A., 16 (72), *21*, 111 (33), 112 (33), *132*
Allingham, M. M., 264 (16), *276*, 295 (32), *298*
Allmand, A. J., 3 (4), *4*
Al-Madfai, S., 209 (31, 32), *217*
Amborski, L. E., 136 (7), 145 (39), *158*, *159*
Amundson, N. R., 284 (18, 19), *287*
Andersen, A. H., 262 (2), *276*
Angelscu, E., 163 (7), *163*
Aniansson, G., 8 (19), *20*
Arakawa, M., 293 (20, 21), *298*
Arakawa, T., 293 (20, 21), *298*
Aranow, R. H., 212 (41), *217*
Argument, C., 10 (45), *20*, 266 (25), *276*
Argyle, A. A., 8 (20), 17 (76), *20*, *21*, 271 (50), *277*
Avgul', V. T., 6 (4), *19*

Bagot, J., 240 (44), *243*
Baker, L. A., 279 (1), *286*
Bakr, A. M., 35 (12), *68*
Bankston, P. T., 292 (12), *298*
Barker, J. T., 8 (16), *20*
Barrow, R. F., 281 (8), *287*

Bartell, F. E., 7 (6), 14 (57), *19*, *21*, 39 (16), 37 (16), *68*, 71 (3), 72 (3), 73 (3), 74 (3), 77 (3, 9), 79 (13), 80 (13), *84*, *85*, 165 (2, 3, 5), 185 (73), *189*, *190*, 197 (8), 203 (8), 207 (8, 24), 208 (24), *216*, *217*, 226 (21), 227 (21), *229*, 253 (15), 255 (22), *260*, *261*
Bartell, F. S., 166 (14), 185 (66), *189*, *190*
Bartell, L. S., 15 (63), *21*, 109 (28), *132*
Baum, A., 182 (57), *190*
Baylé, G. C., 286 (33, 34), *287*
Baylé, G. E., 27 (12), *31*
Bayley, C. H., 267 (29), *277*
Belton, J. W., 247 (6), *260*
Benesi, H., 294 (30), *298*
Benner, F. C., 7 (6), *19*, 165 (2), *189*, 197 (8), 203 (8), 207 (8), *216*
Berkowitz, B. J., 14 (59), *21*
Bernal, J. D., 23 (2), *31*
Berthier, P., 18 (86), *21*
Besson, P., 294 (29), *298*
Beyer, G. L., 152 (55), *159*
Bigelow, W. C., 116 (47), *132*
Bikerman, J. J., 9 (33), *20*
Binsford, J. S., 136 (13), 140 (13), 150 (13), *158*
Birrell, K. S., 292 (13), *298*
Bit-Popova, E. S., 223 (14), *229*
Black, C. E., 136 (7), *158*
Black, W., 267 (26), *277*
Blackburn, A., 18 (89, 91), 19 (89), *22*, 33 (8, 10), 44 (8), 46 (8), 47 (8), 55 (8), 56 (8), 59 (8), *68*, 176 (39, 40), 182 (60), 183 (60), 184 (60), *190*, 255 (23), *261*
Blacker, H. G., 291 (7), *297*
Blair, C. M., 238 (39), *243*
Blake, F. C., 127 (59), *132*
Bloor, W. R., 15 (61), *21*

299

INDEX OF SYSTEMS

(In addition to the specific systems which are indexed, a number of general references occur in the text, especially in Chapter 17.)

311

formic acid—carbon tetrachloride, 81, 83
glutaric acid—water, 124
n-heptanoic acid—water, 185–186
n-hexatriacontane—carbon tetrachloride, 119
iodine—cyclohexane, 102
iodine—various organic solvents, 128–129
malonic acid—water, 99, 124
methyl alcohol—benzene, 65, 166
methylene blue—water, 295
n-octacosane—carbon tetrachloride, 119
n-octadecane—carbon tetrachloride, 119
n-octane—methyl alcohol, 74
oxalic acid—water, 124
pimelic acid—water, 124
polyisobutene—benzene, 140
polyisobutene—carbon tetrachloride, 140
polyisobutene—di-isobutene, 143
propionic acid—water, 185–186
sebacic acid—water, 124
stearic acid—benzene, 97
stearic acid—carbon tetrachloride, 97
stearic acid—cyclohexane, 97
stearic acid—ethyl alcohol, 97
succinic acid—water, 124
n-valeric acid—water, 185–186

Spheron 6 (1000°c)

n-butanol—water, 77
n-butyric acid—cyclohexane, 55–56
chlorine—carbon tetrachloride, 88–89
n-propanol—water, 54

Spheron 6 (1400°c)

chlorine—carbon tetrachloride, 88–89

Spheron 6 (2700°c), *Graphon*

acetic acid—water, 74, 76
adipic acid—water, 121–124
benzoic acid—water, 100–101, 232
bromine—carbon tetrachloride, 66–67
butyl alcohol—benzene, 173, 175
n-butyl alcohol—water, 74, 76, 77, 230, 258–259
n-butyric acid—water, 74, 76
n-caproic acid—water, 74, 76
chlorine—carbon tetrachloride, 88–89
cyclopentanone—n-heptane, 50, 51
n-decane—methyl alcohol, 74–75
n-docosane—carbon tetrachloride, 118–119
n-dodecane—methyl alcohol, 74
n-dotriacontane—carbon tetrachloride, 118–119

polystyrene—toluene, 150–151
polyvinyl acetate—toluene, 136, 231
propionic acid—water, 180–185
propyl alcohol—benzene, 173–174, 177
iso-propylbenzene—cyclohexane, 120, 167
pyridine—ethyl alcohol, 176
pyridine—water, 47, 55–56, 176
salicylic acid—water, 170
stearic acid—benzene, 25
succinic acid—water, 24, 91, 121–122, 170
triethylamine—water, 74–75, 78–79
n-valeric acid—water, 73, 181, 184–185
o-xylene—cyclohexane, 120, 167
p-xylene—cyclohexane, 120, 167
acetic acid—acetone—water, 160
aniline—acetic acid—water, 163
aniline—butyric acid—water, 163
aniline—propionic acid—water, 163
oxalic acid—succinic acid—water, 163
toluidine—acetic acid—water, 163
toluidine—butyric acid—water, 163
toluidine—propionic acid—water, 163

GLASS

n-eicosyl alcohol—hexadecane, 116
iodine—carbon tetrachloride, 128, 294
hexamethyl disiloxane—benzene, 140
n-nonadecanoic acid—hexadecane, 116
n-octadecylamine—hexadecane, 116
perfluoro-octanoic acid—n-decane, 232
polydimethylsiloxane—benzene, 136
polyesters—chloroform, 145
polyesters—toluene, 145
polyethylene glycol—benzene, 157
polyethylene glycol—methyl alcohol, 157
polystyrene—toluene, 145

GRAPHITE

acetic acid—water, 71
n-butyl alcohol—water, 79–80, 207
n-butyric acid—water, 71
n-caproic acid—water, 71, 72–73
cyclohexane—benzene, 167–168
n-decane—methyl alcohol, 74
n-dodecane—methyl alcohol, 74
n-heptylic acid—water, 71
hexadecyl sulphate—water, 269
methylene blue—water, 295
octadecyl sulphate—water, 269
n-octane—methyl alcohol, 74

LIQUID–VAPOUR

n-nonadecanoic acid—hexadecane, 116
n-octadecylamine—hexadecane, 116
palmitic acid—benzene, 103
stearic acid—benzene, 105, 113

Silver

ethyl stearate—benzene, 117
n-heptylic acid—water, 7
n-nonadecanoic acid—cyclohexane, 112
octadecyl alcohol—benzene, 107–108, 113, 117
n-octanoic acid, 7
stearic acid—benzene, 105, 113

Steel

capric acid—benzene, 103
n-cetylamine—nujol, 93
dimethyl-n-cetylamine—nujol, 93
stearic acid—benzene, 103

Tin

polyvinyl acetate—benzene, 136, 146
polyvinyl acetate—carbon tetrachloride, 136, 146
polyvinyl acetate—dichlorethane, 136, 146

Zinc

aliphatic alcohols—water, 7
stearic acid—benzene, 105

MOLECULAR SIEVES

benzene—n-hexane, 43, 167–168
n-butyl chloride—benzene, 43
n-butyl chloride—carbon tetrachloride, 43
furfuryl alcohol—water, 43
polydimethylsiloxane—hexane, 143

SILICA AND SILICA GEL

acetic acid—carbon tetrachloride, 170
acetic acid—heptane, 78–79
iso-amyl alcohol—iso-octane, 197
n-butyl alcohol—benzene, 34
carbon tetrachloride—benzene, 249
crystal violet—water, 264
cyclohexanol—iso-octane, 197
dichloracetic acid—water, 230
2,4-dimethylpentane—benzene, 52
ethyl alcohol—benzene, 54–55, 61, 66, 164–165
ethyl alcohol—water, 177
ethylene dichloride—benzene, 59, 62, 168
ethylene oxide condensates—water, 144, 154–155
fatty acids—toluene, 185

SUBJECT INDEX

Adhesion, 134
Adhesion tension, 7
Adsorbed layer
  number of molecular layers in, 52, 54, 145, 268–269
  thickness of, 48–49, 52, 197–201, 203, 260
Adsorption, co-operative, 131
  see also Lateral interactions
  endothermic, *see* Endothermic adsorption
Adsorption from mixed vapours, 11, 12, 24, 32–35, 44, 52
Adsorption from solution
  absolute, 27
  apparent, 27
  chemisorption, 2, 62–67, 81–84, 87–89, 90, 102–106, 111–112, 115, 128–129, 130, 283, 289, 292
  competitive, 164, 272
  individual, 27
  irreversible, 116, 128–129
  "negative," 26, 27, 30, 140
  physical, 2
  preferential, 28, 29, 164, 165–168, 173, 175–176, 178, 195, 201, 252, 254, 279
  reversible, 113
  selective, 27, 164
Adsorption from solution
  decomposition of adsorbate, 19, 282
  dissolution of adsorbent, 19, 104, 107, 111, 292
  effect of out-gassing adsorbent, 18, 87
  effect of pressure, 188–189
  effect of temperature, 60–61, 79–80, 101–102, 117, 188, 221, 223, 234
  kinetics of, 2, 92–93
  measurement at elevated temperature, 5, 6
  reaction catalysed by adsorbent, 19, 81, 83, 86–88, 283
Adsorption isotherm, *see* Isotherm
Amagat's equation, two-dimensional analogue of, 198

Analysis of solutions
  colorimetric, 14, 15, 294
  densitometric, 16
  gravimetric, 15, 16
  miscellaneous methods, 17
  refractometric, 13, 14
  titrimetric, 15
  using radioactive tracers, 16, 17, 112, 232, 271, 292
  using spread films, 17
Area-filling, 42, 169
Atropic mixture, 162
Autophobic films, 116
Avogadro's number, 41

Barrier method, McBain's, 7, 191
Breakdown time, 279–283
Breakdown volume, 280
Bunsen coefficient, 86

Chromatographic analysis, 6, 294
Chromatography
  gas–liquid, 8, 279
  liquid–liquid, 279
  liquid–solid, 160, 172, 188, 283–286
  use in separations, 138, 279
Columns
  adsorption from, 10
  adsorption on, 2, 86, 188, 259, 279–286
Critical micelle concentration, 155, 263, 267, 268, 269, 272
Critical miscibility temperature (polymers), *see* Theta temperature
Critical solution temperature, 77–79, 188

Desorption, 112, 119, 152–154, 234, 281–282
Detergents, adsorption of, 3, 10, 262–263, 266–275
Dipole moment, 179
Displacement, 17
Displacement development, 283, 286